A FORMULARY OF DETERGENTS
AND OTHER CLEANING AGENTS

Compiled by

Michael and Irene Ash

Chemical Publishing Co.
New York, N.Y.

PREFACE

Through the cooperation of the contributors of the major surfactant-producing corporations, we are pleased to present a formulary devoted entirely to a variety of cleaning agents. The purpose of this compilation is not to educate the reader as to the physical and chemical nature of detergent ingredients but to make available a catalogue of formulas reflecting the current technology in the surfactant industry.

The formulas herein should be considered starting-point preparations and thereby used as the basis for further experimentation in the development of new detergent products.

The book is organized so that each chapter deals with major detergent applications. Within each chapter the formulas are further subdivided according to their hydrophilic nature which determines their mode of activity. These four categories are: anionic, nonionic, cationic, and amphoteric.

Unless otherwise specified, all formulas have the quantities of ingredients given in parts by weight. A list of abbreviations that are used throughout the formulary is included. All constituents appearing by their tradename are printed in boldface type, and the manufacturers' names and addresses appear after the list of alphabetized tradenames in the appendix.

ABBREVIATIONS

@ . . . . at
approx. . . . . approximately
aq. . . . . aqueous
Be . . . . Baume
C . . . . degrees Centigrade
cc. . . . . .cubic centimer
cm . . . . .centimer
conc.. . . . . .concentrated
cps . . . . centerpoise(s)
cs. . . . . .centistoke
F . . . . degrees Fahrenheit
fl. oz. . . . . fluid ounce
ft . . . . .foot
g . . . . gram
gal . . . . .gallon
h . . . . hour
ht. . . . . height
in. . . . . .inch
l. . . . . .liter
M. . . . . mole
max . . . . .maximum
med. . . . . medium
min. . . . . .minute
min. . . . . minimum
ml . . . . .milliliter
mm. . . . . .millimeter
m.p. . . . . .melting point
neut.. . . . . neutralized
NF . . . . National Formulary
No.. . . . . number
O/W . . . . oil in water
oz. . . . . .ounce
POE . . . . polyoxyethylene
ppm . . . . .parts per million
q.s. . . . . quantity sufficient to make
rpm . . . . revolutions per minute

ABBREVIATIONS

s. . . . . second
sol'n. . . . . solution
Tw. . . . . Twaddell
USP . . . . United States Pharmacopeia
visc. . . . . viscosity
W/O . . . . water in oil
wt. . . . . weight
XXX. . . . . triple pressed
≈ . . . . approximately
. . . . number
% . . . . percent
qt. . . . . quart

CONTRIBUTORS

Alkaril Chemical Ltd.
3265 Wolfedale Road
Mississauga, Ont., Canada

Amerchol Corp.
Unit of CPC Internat'l Inc.
Amerchol Park
Edison, N.J. 08817

Armak Chemicals Div.
Box 1805
Chicago, Ill. 60690

Ashland Chemicals
Chemical Products Div.
Box 2219
Columbus, Ohio 43216

Clintwood Chemical Co.
4342 South Wolcott Ave.
Chicago, Ill. 60609

Emulsion Systems, Inc.
215 Kent Ave.
B'klyn., N.Y. 11211

General Electric Co.
Silicone Products Dept.
Waterford, N.Y. 12188

Henkel Inc.
1301 Jefferson St.
Hoboken, N.J. 07030

Inolex Corp.
Personal Care Div.
4221 S. Western Blvd.
Chicago, Ill. 60609

ICI Petrochemicals Div.
PO Box 90
Wilton, Middlesborough
Cleveland, TS6 85E, England

Mazer Chem. Inc.
3938 Porett Dr.
Gurnee, Ill. 60031

Mona Industries
176 E. 24th St.
Paterson, N.J. 07524

Patco Cosmetic Prod.
C.J. Patterson Co.
3947 Broadway
Kansas City, Mo. 64111

Pennwalt Corporation
900 First Ave.
King of Prussia, Pa. 19406

Stepan Chem. Co.
Edens and Winnetka
Northfield, Ill. 60093

Tomah Products, Inc.
1012 Terra Drive
PO Box 388
Milton, Wisc. 53564

R.T. Vanderbilt Co., Inc.
30 Winfield St.
Norwalk, CT 06855

Witco Chem. Corp.
Ultra Div.
2 Wood St.
Paterson, N.J. 07524

TABLE OF CONTENTS

Chapter I

HYGIENIC CLEANERS

Bubble Bath

	Formula No. 1	No. 2 (Anionic)	No. 3
Alkasurf T	–	–	20
Alkasurf ES 60	–	5	–
Alkasurf EA 60	16	–	–
Alkasurf ALS	8	10	10
Alkamide CDE	4	3	3
Water	to 100	to 100	to 100
Phosphoric Acid	to pH 7	to pH 7	to pH 7
Preservative	q.s.	q.s.	q.s.
Sodium Chloride	0.5	1.0	1.0
Color and Perfume	q.s.	q.s.	q.s.

Procedure:
While agitating the water, add the ingredients in the above order.

	No. 4	No. 5 (Anionic)	No. 6
Alkasurf ES 60	–	–	23
Alkasurf ES 30	–	20	–
Alkasurf T	30	–	–
Alkamide L9DE	10	3	–
Alkamide L7DE	–	–	4.5

1

Water	60	76	71.5
Citric Acid	←	to pH 7	→
Sodium Chloride	–	1-2	1-2
Preservative	q.s.	q.s.	q.s.
Color and Perfume	q.s.	q.s.	q.s.

Note:
A sparkling clear product can be produced that can be readily perfumed. Should a more economical formulation be required, use **Alkamide CL63**.

	No. 7	No. 8
		(Anionic)
Neodol® 25-3S (60% AM)	32	32
Lauryldimethylamine Oxide (30% AM)	16	16
Total Actives (100% AM)	24	24
Bubble Bath Fragrance	1	–
Shampoo Fragrance	–	0.3
FD&C Blue No. 1 (1% aq. sol'n.)	0.1	0.1
Water	q.s.	q.s.

No. 9

(Nonionic-Anionic)

A **Bio Terge AS-40**		40.0
Pationic ISL		3.0
Monamid 716		3.0
Perfume		1.0
B **Glydant (40-700)**		0.2
Water (deionized)		52.8
pH	7.3	
Cloud Point	<–3 C	
Clear Point	–	

Procedure:
Combine ingredients of Part A. Heating gently will hasten solubility.

Combine ingredients of Part B. Add B to A with agitation. Cool to room temperature if necessary.

	No. 10	No. 11
	(Nonionic-Anionic)	
Alkasurf ALS	10	—
Alkasurf SS LA3	—	15
Alkasurf ES 60	10	15
Alkamide CDO	4	5
Soluble Lanolin	—	2
Sodium Chloride	1.0	1.0
Water	to 100	to 100
Color and Perfume	q.s.	q.s.

Note:

The above formulations contain free glycerin which contributes to the emollient properties of the finished product. The use of the **Alkasurf** sulfosuccinate surfactants offers improved soap and hard water resistance in addition to extra mildness.

No. 12

(Nonionic-Anionic)

Water	84.5
Sodium Chloride	2.5
Standapol® WAQ Spec.	2.0
Standamid® SD (Cocamide DEA)	2.0
Propylene Glycol	1.0
Standapol® ES-40 Conc. (Sodium Myreth Sulfate)	8.0
Perfume Oil	q.s.
Dyes and Preservatives	q.s.

Procedure:

The order of addition is given above. Add all materials singly under adequate agitation. Continue stirring until product is homogeneous.

Note:
This low actives (9%) formula provides copious foam in the tub.

No. 13

(Nonionic-Anionic)

Water	68.5
Sodium Chloride	1.0
Standapol® ES-40 Conc.	20.0
Standamid® LD (premelted at 45 C)	5.0
Standamul® HE	2.5
Standamox® CAW	3.0
Perfume Oil	q.s.
Dyes and Perservatives	q.s.

Procedure:
The order of addition is given above. Add all materials singly under adequate agitation. Continue stirring until product is homogeneous.

Note:
The ethoxylated cocoate provides emolliency. The blend of amide and amine oxide provides a high level of detergency with minimum irritation potential not possible in a strictly high amide blend.

Herbal Bubble Bath

Formula No. 1

(Nonionic-Anionic)

Water	50.0
Sodium Chloride	2.0
Standapol® WAS-100	20.0
Standapol® ES-7099	20.0
Standamid® SD	2.0
Standamul® OXL	3.0
Sedaplant Richter®	3.0
Perfume Oil	q.s.
Dyes and Preservatives	q.s.

Procedure:
The order of addition is given above. Add all ingredients singly under adequate agitation. Continue stirring until product is homogeneous.

Note:
The blend of propoxylated-ethoxylated fatty alcohol and CLR material provides emollient and substantive dermal effects.

No. 2

(Amphoteric)

Lexaine IBC-70	20.00
Sodium Chloride	0.10
Lexein-X250	1.00
Bronopol	0.02
FD&C Yellow No. 5 (½%)	0.80
FD&C Blue No. 1 (½%)	0.08
FD&C Red No. 4 (½%)	0.04
Perfume O-115 (UOP)	0.50
Water	77.38

Visc.: Brookfield RVT (#3, 20 rpm, 1 min) = 2000 cps, pH = 4.95

Procedure:
Charge water into a suitable making tank equipped with an agitator and having provisions for heating and cooling. Heat to 70-90 C with moderate agitation. Increase agitator speed and gradually add **Lexaine IBC-70**. When this is all added and completely dissolved, cool with moderate agitation to 40-45 C; then add and disperse balance of ingredients in the order shown. Cool to 25 C and fill.

Emollient Bubble Bath

(Nonionic-Anionic)

Water	29.0
Standapol® ES-2	60.0
Standamul® HE	5.0
Standamid® LD (premelted at 45 C)	3.0

Perfume Oil	3.0
Dyes and Preservatives	q.s.

Procedure:

The recommended order of addition is given above. Add all materials singly under adequate agitation. Continue stirring until product is homogenous.

Note:

The ethoxylated cocoate provides emollient and substantive dermal effects.

Low Irritation Shampoo

Formula No. 1

(Nonionic-Anionic-Amphoteric, Conditioning)

Water	48.0
Sodium Chloride	1.0
Standapol® BAW	12.0
Standapol® 130-E	12.0
Standapol® ES-2	20.0
Standamid® SD	2.0
Polyquart® H-7102	5.0
Perfume Oil	q.s.
Dyes and Preservatives	q.s.

Procedure:

The order of addition is given above. Add all materials singly under adequate agitation. Adjust the pH to 7.0 ± 0.5 with 50 percent citric acid aqueous solution. Continue stirring until product is homogeneous.

Note:

Dermal and ocular irritation data available upon request. The blend of betaine and ethoxylated sulfate contributes to low irritation potential. The quaternary provides substantivity and conditioning to the hair shaft.

No. 2

(Nonionic-Anionic-Amphoteric)

Water	45.0
Sodium Chloride	1.0
Standapol® ES-2	20.0
Standapol® 130-E	12.0
Standapol® BAW	15.0
Standamid® SD	2.0
Standamox® CAW	3.0
Standamul® HE	2.0
Perfume Oil	q.s.
Dyes and Preservatives	q.s.

Procedure:
    The order of addition is given above. Add all materials singly under adequate agitation. Adjust the pH to 7.0 ± 0.5 with 50 percent citric acid aqueous solution. Continue stirring until product is homogeneous.

Note:
    Dermal and ocular irritation data available upon request. The blend of ethoxylated sulfate, betaine, amine oxide and ethoxylated cocoate contributes to low irritation potential while also providing effective detergency.

No. 3

(Nonionic-Anionic-Amphoteric)

Water	56.0
Standapol® AB-45	20.0
Standapol® ES-2	20.0
Standamid® SD	2.0
Standamul® HE	2.0
Perfume Oil	q.s.
Dyes and Preservatives	q.s.

Procedure:
    The order of addition is given above. Add all materials singly under adequate agitation. Adjust the pH to 7.0 ± 0.5 with 50 percent citric acid

aqueous solution. Continue stirring until product is homogeneous.

Note:
Dermal and ocular irritation data available upon request. The blend of betaine, anionic, and ethoxylated cocoate contributes to low irritation potential while also providing effective detergency.

No. 4

(Anionic)

Alkateric 2CIB	10-30
Alkasurf ES-60	10-25
Alkamox CAPO	5-10
Preservative	q.s.
Color and Perfume	q.s.
Water	to 100

No. 5

(Anionic)

Alkateric BC	10-30
Alkasurf ES-60	10-25
Alkamide CDE	5-10
Preservative	q.s.
Color and Perfume	q.s.
Water	to 100

No. 6

Lexaine IBC-70	10.00
Ninol 2012 Extra	3.00
Lexgard M	0.15
Lexgard P	0.05
Citric Acid Monohydrate	0.40
Sodium Chloride	0.20
Plurafac C-17	1.00
FD&C Yellow No. 5 (½%)	0.20

FD&C Red No. 4 (½%)	0.03
Perfume 802169 U	0.05
Water	84.92

Visc.: Brookfield RVT (#3, 20 RPM, 1 min) = 475 cps (may be increased by reducing **Plurafac** or increasing sodium chloride content)
pH = 7.05 (may be varied by adjusting citric acid content)

Procedure:
Charge water into a suitable making tank equipped with an agitator and having provisions for heating and cooling. Heat to 85-90 C with moderate agitation. Increase agitator speed and gradually add **Lexaine IBC-70**. When this is all added and completely dissolved, add balance of ingredients except colors and perfume. Cool with continued stirring to 40-45 C, then disperse colors and perfume. Cool to 25 C and fill.

Concentrated Shampoo

Formula No. 1

(Nonionic-Amphoteric)

Ammonium Laureth Sulfate (60%)	15.00
Ammonium Laurylsulfate (30%)	20.00
Cocamide DEA	16.00
Lexaine C	12.00
Ammonium Chloride	4.00
Propylene Glycol	8.00
Lexgard M	0.15
Lexgard P	0.05
Bronopol	0.05
Phosphoric Acid to pH = 6.0	q.s.
Perfume CS 18480[1]	0.25
Water, Dye	q.s. to 100.00

1:7 Dilutable

---
[1] Albert Verley

No. 2

(Nonionic-Anionic, Low Irritation)

Standapol® SHC-101	72.25
Propylene Glycol	7.20
Sodium Chloride	1.10
Standamid® LD (premelted at 45 C)	17.00
Citric Acid (50% aq. sol'n.)	1.25
Water	1.20
Perfume Oil	q.s.
Dyes and Preservatives	q.s.

Procedure:

The order of addition is given above. Add all materials singly under adequate agitation. Continue stirring until product is uniform.

Note:

Dermal and ocular irritation data available upon request. The sulfosuccinate-anionic blend inhibits irritation potential while providing synergistic detergency in this liquid concentrate.

No. 3

(Nonionic-Anionic)

Professional shampoos that can be diluted 8 oz to one gal (1:19) can be composed as follows:

Alkasurf DLS	47
Alkamide L9DE	46
Sodium Sulfate	2-3
Denatured Alcohol	5
Preservative	q.s.
Citric Acid	to pH 7
Color and Perfume	q.s.

No. 4

(Nonionic-Anionic)

A	Water	19.25
	Sodium Chloride	7.50
	Propylene Glycol	8.00
	Standapol® TS-100	20.00
	Standamid® SD	12.50
	Standamid® LD (premelted at 45 C)	12.50
	Citric Acid (anhydrous)	0.75
	Standapol® EA-40 Conc.	19.50
B	Perfume Oil	q.s.
	Dyes and Preservatives	q.s.

Procedure:

Heat water to 50-55 C. Keep temperature constant. Add remaining ingredients of A, one at a time, in order listed under agitation. Cool, and at 40-45 C add individual components of B under agitation. Adjust the pH to 6.5 ± 0.5 with 50% citric acid aqueous solution. Continue sweep-type agitation until product reaches room temperature. Finished shampoo can be made by diluting 1 part concentrate with 7-15 parts water.

Note:

This liquid high actives (60 percent) blend provides excellent detergency even after dilution.

No. 5

(Nonionic-Anionic, Beauty Parlor)

A	**Maprofix DLS-35**	46.0
	Methyl Paraben	0.1
	Water (deionized)	q.s. to 100.0
B	**Ninol 2012E**	35.0
	Kessco® PEG 6000 Distearate	2.0
C	Citric Acid (pH adjust)	1.0
	Perfume and Color	as desired

Procedure:
Blend A heating to 60 C. Melt B and add to batch heating to 75 C.
Mix 15 min. Cool to 40 C. Add C and cool to 30 C.

No. 6

Lexaine IBC-70	30.00
Sodium Chloride	0.70
Lexgard M	0.15
Lexgard P	0.05
Sodium Hydroxide (25%)	0.40
D&C Green No. 8 (½%)	0.50
FD&C Blue No. 1 (½%)	0.06
Perfume # 8240[1]	0.20
Water	67.94

[1] Norda

Visc.: Brookfield RVT (#6, 10 RPM, 1 min) = 31,000 cps
pH = 7.28 (pH must be above 7.0 for D & C Green No. 8 to
exhibit dichroism that adds brightness to product.)

Procedure:
Charge water into suitable making tank equipped with an agitator and
having provisions for heating and cooling. Heat to 70-90 C with moderate
agitation.   Increase agitator speed and gradually add **Lexaine IBC-70**.
When this is added and completely dissolved, add and dissolve sodium chloride and **Lexgards**. Cool to 45-50 C; then add and disperse colors and perfume. Adjust pH with sodium hydroxide after determining amount required to just exceed 7.0 pH. Fill at 40-45 C into tubes, or alternatively cool to 25 C and fill depending on capacity and design of filling equipment.

Low pH Shampoo

Formula No. 1

(Nonionic-Anionic-Amphoteric)

Water	58.0

Propylene Glycol	2.0
Standapol® A	30.0
Standapol® BAW	5.0
Standamid® LDM	2.0
Standamox® CAW	3.0
Perfume Oil	q.s.
Dyes and Preservatives	q.s.

Procedure:

The order of addition is given above. Add all materials singly under adequate agitation. Adjust the ´pH to 5.0±0.5 with 50 percent citric acid aqueous solution. Continue stirring until product is homogeneous.

Note:

The ammonium anionic in combination with the betaine and amine oxide retain low pH in this system.

No. 2

(Anionic-Amphoteric, Lotion)

Water	57.7
Magnesium Lauryl Sulfate (27%)	29.6
Monamate CPA (40%)	5.0
Monateric ISA (35%)	5.7
Cersynt IP	2.0

Visc.: approximately 12,000–15,000 cps
pH "as is" 5.5

Procedure:

Add components in above order. Warm to 50 C with slow agitation.

Acid-Conditioning Shampoo

(Nonionic-Anionic)

A	**Standapol A**	53.0
	Clindrol 100 LM	3.0
	Pationic ISL	3.0
	Perfume A-7131	0.3

B Methyl Paraben	0.2
Water (deionized)	40.5
Lactic Acid (44%)	q.s.
Cloud Point	8 C
Clear Point	10 C
Visc.: (Brookfield RVT–Spindle #2 @	
20 rpm @ 80 F	650 cps

Procedure:
Combine ingredients of Part A. Heating gently will hasten solubility. Combine ingredients of Part B. Add Part B to A with agitation. Adjust pH to ± 5.5 with lactic acid.

Pearly Shampoo

Formula No. 1

(Nonionic-Amphoteric)

Sodium Laurylsulfate (30%)	30.00
Lexaine C	6.00
Lexemul EGMS	1.00
Sodium Chloride	1.40
Compound 213.008[1]	0.20
Water, Dye, Preservative	q.s. to 100.00

---

[1] Firmenich

Procedure:
Add sodium laurylsulfate, **Lexaine C, Lexemul EGMS** and sodium chloride to water and heat to 65-70 C. When the **EGMS** is completely dispersed, cool to 45 C. Adjust pH to 6.0-6.5 with phosphoric acid, and add dye, preservative and perfume. Fill at 35 C.

No. 2

(Nonionic-Anionic)

| **Alkasurf WAQ** | 30 |

Alkamide L9DE	4
Ethylene Glycol Monostearate	2-3
Water	63
Sodium Chloride	1
Citric Acid	to pH 7
Preservative	q.s.
Color and Perfume	q.s.

Procedure:

To approximately half of the required water, melt together the EGMS, **Alkasurf WAQ, Alkamide L9DE**, taking care to obtain a clear mix. On slowly adding the balance of the cold water, a white pearly shampoo results. Add sodium chloride and preservative and adjust to pH 7.

No. 3

(Nonionic-Anionic)

Alkasurf WAQ	50
Alkamide L9DE	6
Sodium Chloride	1
Water	43
Preservative	q.s.
Citric Acid	to pH 7
Color and Perfume	q.s.

Gel Shampoo

Formula No. 1

(Amphoteric, Economy)

Ammonium Laurylsulfate (30%)	25.00
Lexaine C	10.00
Sodium Chloride	1.50
Lexgard M	0.15
Lexgard P	0.05
Compound 67.079B[1]	0.15

Phosphoric Acid to pH 5.0   q.s.
Water, Dye   q.s. to 100.00

---
[1] Firmenich

Procedure:
   Charge components and heat to 50 C. Adjust pH with phosphoric acid and stir slowly until clear. Add perfume at 45 C and fill at 35 C or less.

No. 2

(Amphoteric, High Active)

TEA Laurylsulfate (40%)	50.00
Lexaine C	15.00
Sodium Chloride	2.50
Lexgard M	0.15
Lexgard P	0.05
Perfume CS 18482[1]	0.15
Water, Dye	q.s. to 100.00

---
[1] Albert Verley

Procedure:
   Charge components to water and heat to 50 C. Stir slowly until clear. Cool to 35 C and fill after product is deaerated.

No. 3

(Anionic–Amphoteric)

Water	36.8
Standapol® ES-2	50.0
Standapol® AB-45	13.2
Perfume Oil	q.s.
Dyes and Preservatives	q.s.

Procedure:
   Blend ingredients in order listed at 55 C, until uniform. Cool to 50 C

with sweep-type agitation to prevent air entrapment.

Note:
The betaine provides emolliency and unique reduction of irritation effects of the anionic in this low actives (18%) system. The betaine also aids in building gel structure.

Protein Shampoo

Formula No. 1

(Nonionic)

A Sodium Lauryl Sulfate (28%)	30.0
Ninol 2012E	6.0
Methyl Paraben	0.2
Super-Pro 5A	3.0
Water (deionized)	q.s. to 100.0
B **Kessco®** PEG 6000 Distearate	2.0
C **Dowicil 200**	0.10
D Citric Acid (pH adjust)	q.s.
Perfume/Color	as desired

Procedure:
Mix A heating to 60 C. Add B heating to 75 C. Mix 15 min, cool to 50 C. Dissolve C in 10 times its weight in water, then add to batch. Add D while cooling to 30 C.

No. 2

(Nonionic–Anionic, Acid-Conditioning)

A **Sulframin 14-16 ADS**	37.5
Pationic ISL	3.0
Clindrol 100 LM	3.0
B Methyl Paraben	0.2
Water (deionized)	53.0

C Perfume A07131   0.3
   Tegin 55 G   3.0
   Lactic Acid (44%)   q.s.

     Visc.: (Brookfield Model RVT Spindle #2 @
        20 rpm @ 80 F)   650 cps
     Cloud Point < -2 C
     Clear Point   —

Procedure:
Combine ingredients of Part A. Heating gently will facilitate solubility.
Combine ingredients of Part B. Add B to A with agitation. Cool to 45 C
and add Part C with continued agitation. Cool to room temperature. Adjust pH to ± 5.5 with lactic acid.

No. 3

(Amphoteric)

Lexaine IBC-70	20.00
Sodium Chloride	0.20
Lexein-X250	5.00
Bronopol	0.02
FD&C Yellow No. 5 (½%)	0.10
Perfume #2478[1]	0.10
Water	74.40

---

[1] Norda

     Visc.: Brookfield RVT (#3, 20 rpm, 1 min) = 1125 cps
     pH = 6.5

Procedure:
Charge water into a suitable making tank equipped with an agitator and
having provisions for heating and cooling. Heat to 70-90 C with moderate
agitation. Increase agitator speed and gradually add **Lexaine IBC-70**. When
this is all added and completely dissolved, cool with moderate agitation to
40-45 C. Then add and disperse balance of ingredients in the order shown.
Cool to 25 C and make final pH adjustment as required; then fill.

No. 4

(Amphoteric)

Ammonium Laurylsulfate (30%)	25.00
Lexaine C	14.00
Lexein X250	5.00
Bronopol	0.05
Lexgard M	0.15
Lexgard P	0.05
Perfume M-45790[1]	0.35
Sodium Chloride	q.s.
Water, Dye	q.s. to 100.00

---

[1] Shaw Mudge

Procedure:
  Add all components, except dye and perfume, to water. Heat to 45 C and stir until clear. Add perfume and dye. Adjust viscosity with sodium chloride, cool and fill.

Egg Shampoo

(Nonionic-Amphoteric)

Sodium Laurylsulfate (30%)	18.00
Lexaine C	6.00
Lauramide DEA	1.50
PEG 6000 Distearate	3.00
Lexgard M	0.15
Lexgard P	0.05
Lexemul EGMS	1.50
Egg Oil[1]	1.00
Sodium Chloride	1.50
Perfume 802169U[2]	0.25
FD&C Yellow No. 5[3]	q.s.
Hydrochloric Acid (conc.)	q.s.
Water	q.s. to 100.00

---

[1] Viobin Corp.

[2] PFW
[3] Kohnstamm

Procedure:

Charge water into kettle and heat to 65-70 C. Add all ingredients except perfume, **Lexaine C**, Egg Oil, HCl, and dye, with moderate agitation, then add **Lexaine C** at 65-70 C with agitation after other ingredients are dissolved. Cool batch with stirring to 45 C and add balance of ingredients, adjusting pH to 6.5-7.0 with HCl. Stir and cool to 30 C, then fill.

Clear Liquid Shampoo

Formula No. 1

(Nonionic-Anionic)

Alkasurf DLS	30
Alkamide L9DE	6
Water	64
Phosphoric Acid or Citric Acid	to pH 7
Preservative	q.s.
Color and Perfume	q.s.

No. 2

(Nonionic-Anionic)

Water	49.0
Sodium Chloride	0.5
Lantox 55	0.5
Standapol® T	30.0
Standamid® SD	5.0
Standapol® ES-40 Conc.	15.0
Perfume Oil	q.s.
Dyes and Preservatives	q.s.

Procedure:

Heat water to 50 C. Add remaining ingredients in order listed one at a time, under agitation. Continue stirring until product is homogeneous.

Note:
This high actives (25%) anionic blend is an efficient foamer and cleanser.

No. 3

(Anionic, High Viscosity)

Alkasurf ES 30	40
Alkamide CDE	3.0
Sodium Chloride	q.s.
Water	to 100
Color and Perfume	q.s.
Citric Acid	to pH 6
Preservative	q.s.

Procedure:
To water at 50–60 C add all of the ingredients and stir until homogeneous. Adjust pH to 7.0.

Super-Conditioning Shampoo

(Cationic-Amphoteric)

Lexaine C	10.00
Lexamine C-13	6.00
Lexamine O-13	2.00
Natrosol 250 HHR	0.70
Bronopol	0.04
Compound 213.008[1]	0.15
Water, Dye	q.s. to 100.00

---

[1] Firmenich

Procedure:
Thoroughly disperse the **Natrosol** and heat to 45 C. After dispersed, add **Lexaine C** and **Lexamines**. Add phosphoric acid to pH 4.5-5.0. Add **Bronopol**, dye and perfume.

Conditioner Shampoo

Formula No. 1

(Anionic)

Lakeway 301-10 (A.O.S. 40%)	18.75
Hamposyl C-30	25.00
Methyl Paraben	0.10
Glydant	0.20
Pationic ISL	5.00
Monamid 716	3.00
PEG-6000 Distearate	2.00
Glucam E-10 (glucose derivative, P.O.E. 10)	3.00
Perfume #D-78-315	0.30
Uvinul 490	0.10
Water (deionized)	42.55
Cert. FD&C Blue No. 1	q.s.

No. 2

(Nonionic)

SLM-40T	28.125
Lakeway 301-10 (A.O.S. 40%)	9.375
Clindrol CG	5.000
Glucam P-10	3.000
Emerest 2355	2.000
Sodium Bicarbonate	0.200
Glydant	0.200
Methyl Paraben	0.100
Perfume	0.200
Water (deionized)	51.800

No. 3

(Nonionic–Anionic, Nonstripping)

A	Water	61.0
	Standapol® ES-2	30.00

Standamul® HE	5.0
Standamid® LD (premelted at 45 C)	4.0
B Perfume Oil	q.s.
Dyes and Preservatives	q.s.

Procedure:

Heat water to 60 C. Add other ingredients of A in above listed order of addition. Cool with sweep-type agitation and at 45 C add individual components of B. Continue low agitation until product is uniform.

Note:

The blend of ethoxylated sulfate and ethoxylated cocoate contributes to the mildness of this preparation. The cocoate in particular provides emollient effects to the hair shaft.

Baby Shampoo

Formula No. 1

(Amphoteric)

Miranol® 2MCAS Modified	35.0
Lauric Diethanolamide (high active)	2.0
Tween 20	1.0
Hamposyl L	1.0
Water	61.0

Visc.: 600 cps
pH: 7.0

Procedure:

Heat together **Miranol® 2MCAS Modified, Tween 20,** and water at 70 C. Add melted **Hamposyl L** and mix until uniform. Do not adjust the pH as this may affect the viscosity.

No. 2

(Amphoteric)

Miranol® 2MCAS Modified	30.00
Hexylene or Propylene Glycol	2.00
Tween 20	1.00

Perfume Oil (approx.)	0.12
Lauric Diethanolamide (high active)	1.00
Water	65.88

No. 3

(Amphoteric)

Water	48.6
Monateric CSH-32	40.0
Monateric ISA-35	11.4

Procedure:
Adjust pH to 6.8–6.9 with phosphoric acid.
Visc. = approximately 600 cps.

No. 4

(Amphoteric)

Lexaine IBC-70	10.00
Lexgard M	0.15
Lexgard P	0.05
Sodium Chloride	0.70
Perfume 802169 U	0.05
Water, Dye	q.s. to 100.00

Procedure:
Charge water into a suitable making tank equipped with an agitator and having provisions for heating and cooling. Heat to 40 C with moderate agitation. Increase agitator speed and gradually add **Lexaine IBC-70**. When this is all added and completely dissolved, add balance of ingredients except colors and perfume. The viscosity may be adjusted with sodium chloride and the pH adjusted to 6.5–7.0 with phosphoric acid or TEA.

No. 5

(Amphoteric, High Active)

| Sodium Laureth-12 Sulfate (60%) | 35.50 |
| **Lexaine C** | 12.00 |

Natrosol 250 HHR	0.60
Lexgard M	0.20
Lexgard P	0.10
Bronopol	0.04
Water, Dye, Perfume	q.s. to 100.00

Procedure:

Disperse **Natrosol** in cold water. Heat to 50 C and stir until completely hydrated. Add the remaining components and stir until everything is dissolved. Adjust pH to 6.5-7.0 with sodium hydroxide or phosphoric acid.

Note:

This formula was found to have very low sting properties and passes the Draize eye irritation test.

No. 6

(Anionic–Amphoteric, 25.8% Active)

Water	12.0
Monamate OPA-30	46.3
Monateric CSH-32	41.7

Visc.: approximately 600 cps

Procedure:

Add ingredients in the order listed and blend with slow agitation. No heat is required. Adjust pH with phosphoric acid to 6.0.

No. 7

Lexaine IBC-70	10.00
Lexgard M	0.15
Lexgard P	0.05
Sodium Chloride	0.7
FD&C Yellow No. 5 (½%)	0.20
FD&C Red (½%)	0.03
Perfume 802169 U	0.05
Water	88.82

Visc.: Brookfield RVT (#3, 20 rpm, 1 min) = 120 cps (may be adjusted by varying sodium chloride content)

pH = 6.3 (may be adjusted to desired pH by acid or alkali addition)

Procedure:

Charge water into a suitable making tank equipped with an agitator and having provisions for heating and cooling. Heat to 70-80 C with moderate agitation. Increase agitator speed and gradually add **Lexaine IBC-70**. When this is all added and completely dissolved, add balance of ingredients except colors and perfume. Cool with continued stirring to 40-45 C, then disperse colors and perfume. Cool to 25 C and fill.

Dandruff Control Shampoo

Formula No. 1

(Nonionic-Anionic)

A	Water	41.9
	Veegum HV	0.5
	Standapol® WAQ Special	50.0
	Ethylene Glycol Monostearate	3.0
B	**Standamid® LD** (premelted at 45 C)	3.0
	Biosulphur Fluid CLR®	1.0
	Sodium Chloride	0.6
	Perfume Oil	q.s.
	Dyes and Preservatives	q.s.

Procedure:

Heat water to 75-80 C. Sprinkle in **Veegum** under agitation. Once **Veegum** is well dispersed add remaining ingredients of A under agitation. Keep temperature constant. When A is completely uniform, cool and continue stirring. At 40 C add individual ingredients of B, one at a time, under agitation. Adjust the pH to 8.0±0.5 with 50 percent citric acid aqueous solution. Continue stirring until product reaches room temperature.

Note:

The amide and EGMS provide sufficient viscosity in combination with **Veegum** to suspend the **CLR** sulfur active ingredient.

No. 2

(Nonionic–Anionic, Cream)

A	Veegum	1.0
	Water	42.2
	Citric Acid	0.4
	Igepon AC-78 (83% solids)	18.0
	Igepon TC-42 (24% solids)	25.2
B	Cetyl Alcohol	1.8
	Glyceryl Monostearate A.S.	5.9
	Solulan 98	3.5
C	Vancide 89RE	2.0

Procedure:

Add the **Veegum** to the water slowly, agitating continually until smooth. Add rest of A, heat to 75 C. Mix B and heat to 80 C. Add B to A, mixing until cool. Add C to a small portion of the cream, disperse thoroughly. Add this concentrate to the remainder of the shampoo. Mix until uniform.

Note:

The final pH should be about 5.0.

		No. 3	No. 4
		(Nonionic–Anionic, Cream)	
A	Veegum	1.0	1.0
	Water	38.9	61.5
	Citric Acid	0.3	q.s.
	Igepon AC-78 (83% solids)	20.0	20.0
	Triton X-200 (28% solids)	27.8	—
	Plurafac C-17	—	5.0
B	Modulan	1.0	1.0
	Cetyl Alcohol	3.0	2.0
	Stearic Acid	6.0	—
	Glyceryl Monostearate A.S.	—	7.5
C	Vancide 89RE	2.0	2.0

Procedure:

Add the **Veegum** to the water slowly, agitating continually until smooth. Add the rest of A and heat to 85 C. Heat B to 90 C. Add B to A, mixing until cool. Add C to a small portion of the mixture and disperse thoroughly. Add this concentrate to the remainder of the mixture. Mix until uniform.

Note:

The final pH should be about 5.0.

No. 5

(Anionic, Lotion)

A	**Veegum K**	3
	Water	40
B	**Vancide 89RE**	2
	Solulan 98	1
	Cocoyl Sarcosine	4
C	**Igepon TC-42** (24% solids)	50

Procedure:

Add the **Veegum K** to the water slowly, agitating continually until smooth. Combine B and stir until the **Vancide 89RE** is dispersed. Add B to C. Mix until uniform. Slowly add A to B and C.

Note:

The final pH should be about 5.0.

		No. 6	**No. 7**
		(Nonionic-Anionic, Cream)	
A	**Veegum**	1.0	1.0
	Water	38.9	61.5
	Citric Acid	0.3	q.s.
	Igepon AC-78 (83% solids)	20.0	20.0
	Triton X-200 (28% solids)	27.8	—
	Plurafac C-17	—	5.0

B	Modulan	1.0	1.0
	Cetyl Alcohol	3.0	2.0
	Stearic Acid	6.0	–
	Glyceryl Monostearate A.S.	–	7.5
C	Vancide 89RE	2.0	2.0

Procedure:

Add the **Veegum** to the water slowly, agitating continually until smooth. Add the rest of A and heat to 85 C. Heat B to 90 C. Add B to A, mixing until cool. Add C to a small portion of the mixture and disperse thoroughly. Add this concentrate to the remainder of the mixture. Mix until uniform.

Note:

The final pH should be about 5.0.

No. 8

(Anionic, Lotion)

A	Veegum K	3
	Water	40
B	Vancide 89RE	2
	Solulan 98	1
	Cocoyl Sarcosine	4
C	Igepon TC-42 (24% solids)	50

Procedure:

Add the **Veegum K** to the water slowly, agitating continually until smooth. Combine B and stir until the **Vancide 89RE** is dispersed. Add B to C. Mix until uniform. Slowly add A to B and C.

Note:

The final pH should be about 5.0.

No. 9

(Amphoteric)

TEA Laurylsulfate (40%)	18.00
Lexaine C	7.00

Lauramide DEA	7.00
PEG 6000 Distearate	2.50
Magnesium Stearate	2.00
Sodium Chloride	1.50
Ottasept Extra	1.50
Lexgard M	0.15
Lexgard P	0.05
Perfume #2478[1]	0.25
Water, Dye	q.s. to 100.00

---

[1] Norda

Procedure:

Wet magnesium stearate with TEA laurylsulfate, ¼ water and lauramide DEA. Heat this slurry and stir to dissolution at 65-70 C. Add remaining water to another vessel and heat to 65-70 C. Add all other ingredients except dyes and perfume with stirring. Cool batch to 40 C and add perfume and dyes. Fill batch at 40 C.

Cream Shampoo

Formula No. 1

(Nonionic-Amphoteric)

Sodium Laurylsulfate (30%)	45.00
Lauramide DEA	3.00
Lexemul EGMS	3.00
Lexaine C	5.00
Lexgard M	0.20
Water	q.s. to 100.00

Procedure:

Combine ingredients excluding the **Lexaine C**. Heat with agitation enough to dissolve the laurylsulfate, melt and dissolve the diethanolamide and melt and completely disperse the **Lexemul EGMS**. Add the **Lexaine C** with agitation. Cool with agitation to about 30 C.

No. 2

A **Neo-Fat® 18-55**	6.0
Cetyl Alcohol	1.0

Sodium Lauryl Sulfate (30%)	40.0
Methyl Paraben	0.15
B Perfume	q.s.
C Sodium Chloride	1.5
Sodium Hydroxide Pellets	0.5
Water (deionized)	q.s. to 100.0

Procedure:

Heat A to 70 C. Heat C to 75 C. Add C to A with agitation and cool to 40 C. Add B mixing thoroughly.

Economy Shampoo

(Amphoteric)

Ammonium Laurylsulfate (30%)	18.00
Lexaine C	5.00
Cocamide DEA	3.00
Sodium Chloride	0.20
Lexgard M	0.15
Perfume #2478[1]	0.10
Water, Dye, Preservative	q.s. to 100.00

[1] Norda

Procedure:

The viscosity of this system can be controlled by adding propylene glycol to thin out the shampoo and sodium chloride can be added to.thicken the shampoo.

Germicidal Shampoo

(Amphoteric)

Miranol® C2M Conc.	25.0–30.0
Brij 30	4.0– 4.0
Quaternary Ammonium Salt Germicide (50%)	1.0– 1.0
Water	70.0–65.0

Brij 30 adds to the viscosity of this formulation.

Note:
Miranol® C2M Conc. lends itself ideally to the formulation of medicated shampoos containing quaternary salt germicides which meet the Eye Irritation Test according to the Draize method.

Acid Balanced Shampoo

Formula No. 1

(Cationic-Amphoteric)

TEA Laurylsulfate (40%)	35.00
Lexaine C	10.00
Lexamine R-13	1.50
Compound 213.008[1]	0.15
Citric Acid to pH 4.0	q.s.
Water, Dye, Preservative	q.s. to 100.00

---

[1] Firmenich

Procedure:
Add components to water and heat to 40 C. Adjust pH with citric acid and stir until clear. Cool and fill.

	No. 2	No. 3	No. 4
Neodol 25-3S (60% AM)	25.0	20.0	12.0
Cocodiethanolamide	2.1	3.0	1.5
Cocoamido Betaine (30%)	25.0	–	1.0
PEG 6000 Distearate	–	–	0.25
Sodium Chloride	–	3.5	2.0
Citric Acid	≈ 0.1	≈ 0.1	≈ 0.3
	3.0	–	–
Water, Dye and Perfume	q.s.	q.s.	q.s.

Properties
Visc., cps, 24 C (76 F)	1720	1475	2080
Clear Point, C	8	4	5
F	47	39	41

Ross Miles Foam Ht
  (0.1%w conc.)
  Initial, mm   160   155   110
  After 5 min, mm   160   155   110

Shampoo for Oily Hair

(Nonionic)

A Sodium Lauryl Sulfate (30%)   27.0
   Methyl Paraben   0.15

   Water (deionized)   q.s. to 100.0

B **Monamide 150LW**   5.0
   Kessco® PEG 6000 Distearate   1.5

C Perfume/Color   as desired

Procedure:
Blend A heating to 60 C. Add B heating to 75 C. Mix 15 min and cool
to 40 C. Add C and cool to 30 C.

Shampoo Utility

(Nonionic)

Lexaine IBC-70   20.00
Ninol 2012 Extra   2.00
Lexgard M   0.15
Lexgard P   0.05
FD&C Yellow No. 5 (½%)   0.20
FD&C Red No. 4 (½%)   0.10
D&C Green No. 5 (½%)   0.03
Perfume G73-144[1]   0.10
Water   75.37
Propylene Glycol   2.00

---

[1] Perry Bros.

Visc.: Brookfield RVT (#3, 50 rpm, 1 min) = 1000 cps
pH = 8.1

Procedure:

Charge water into a suitable making tank equipped with an agitator and having provisions for heating and cooling. Heat to 70-90 C with moderate agitation. Increase agitation speed and gradually add **Lexaine IBC-70**. When this is added and completely dissolved, add **Ninol** and **Lexgards** and dissolve. Cool to 40–45 C and add and disperse balance of ingredients. Cool to 25 C and fill after completing viscosity adjustment with propylene glycol.

Shampoo for Ulotrichous Hair

(Nonionic-Anionic)

A	**Maprofix DLS 35**	46.0
	Methyl Paraben	0.2
	Water (deionized)	q.s. to 100.0
B	**Ninol 2012E**	35.0
	Kessco® PEG 6000 Distearate	2.0
C	Citric Acid	1.0
	Wilson's WSP-250	1.0
D	**Dowicil 200**	0.2
E	Perfume and Color	as desired

Procedure:

Blend A heating to 60 C. Melt B and add to batch heating to 75 C. Mix 15 min. Cool batch to 50 C and add C. Dissolve D in 10 times its weight in water and add to batch. Cool to 40 C. Add E and cool to 30 C.

Pearlescent Wig Shampoo

(Nonionic-Anionic-Cationic)

Maprofix ES	5.1
Triton X-100	10.0
Polysorbate 80	7.0
Aromox® C/12W (40%)	5.0

Kessco® Ethylene Glycol Monostearate	2.0
Ethomeen® T/15	2.5
Methyl Paraben	0.2
Water (deionized)	q.s. to 100.0
Color and Perfume	

Procedure:
Melt **Kessco** ethylene glycol monostearate to 65 C. Add remaining agents to blend at 65 C.

Cream Rinse

Formula No. 1

(Cationic, Economy, Opaque)

Lexate CRC	3.40
PEG 600 Distearate	0.30
Lexgard M	0.15
Lexgard P	0.05
Sodium Chloride	0.60
Citric Acid Monohydrate	0.35
Perfume #2478[1]	0.10
Water	95.05
D&C Red No. 19	q.s.

[1] Norda

Visc.: Brookfield RVT (#4, 20 rpm, 1 min) = 1750 cps
pH = 5.3.

No. 2

(Cationic-Amphoteric, Protein)

Lexate CRC	5.50
Stearyl Stearate	1.00
Lexein X250	3.50
Lexgard M	0.15
Lexgard P	0.05
Bronopol	0.05

Citric Acid Monohydrate	0.50
Sodium Chloride	1.00
Perfume 802 169U[1]	0.25
Water	88.00
FD&C Yellow No. 5	q.s.

[1] PFW

Visc.: Brookfield RVT (#4, 20 rpm, 1 min) = 1700 cps
pH = 5.6.

Procedure:
Charge water into making tank and heat to 70-75 C. Dissolve **Lexgards** and citric acid and disperse **Lexate CRC** and stearyl stearate with moderate agitation to avoid incorporating air. Cool to 45 C and add **Bronopol**, perfume, **Lexein**, sodium chloride and color with good agitation. Continue stirring and cool to 25-30 C. Pack into suitable containers and let stand. Viscosity develops fully after 24 h at 25 C.

No. 3

(Economy, Pearlescent)

Lexate CRC	3.40
PEG 600 Distearate	1.00
Lexgard M	0.15
Lexgard P	0.05
Sodium Chloride	0.60
Citric Acid Monohydrate	0.35
Perfume #2478[1]	0.10
Water	94.35
D&C Red No. 19	q.s.

[1] Norda

Visc.: Brookfield RVT (#4, 20 rpm, 1 min) = 1400 cps
pH = 5.4.

Procedure:
Charge water and heat to 70-75 C. Add all ingredients except color,

sodium chloride and perfume. Agitate until uniformly dispersed. Cool with agitation to 50 C and add sodium chloride. Cool with agitation to 45 C and add balance of ingredients. Cool to 30 C and fill.

Hair Conditioner

Formula No. 1

(Amphoteric, Cream)

Lexate CRC	5.50
Stearyl Stearate	1.00
Lexein X250	5.00
Mineral Oil 125/135	2.00
PEG 400 Distearate	1.50
Formalin	0.10
Citric Acid Monohydrate	0.50
Sodium Chloride	1.20
Perfume G73-146[1]	0.25
Water	82.95

---

[1] Perry Bros.

Consistency: after 24 h @ 25 C—soft cream; pH = 5.5.

Procedure:

Charge water into making tank and heat to 70-75 C. Dissolve citric acid and sodium chloride; then add and disperse **Lexate CRC**, stearyl stearate, mineral oil and PEG 400 distearate. Continue stirring and cool to 50-55 C, and add and disperse **Lexein,** formalin, and perfume. Continue cooling with gentle agitation to 35-40 C. Fill into suitable containers. Consistency develops fully after 24 h at 25 C.

No. 2

(Amphoteric, Lotion)

Lexate CRC	5.50
Stearyl Stearate	1.00
Lexol IPP	1.00

Hexylene Glycol	5.00
Formalin	0.10
Citric Acid Monohydrate	0.50
Sodium Chloride	1.00
Perfume M-45790[1]	0.25
Water	85.65
D&C Green No. 5	q.s.
D&C Yellow No. 10	q.s.

[1] Shaw Mudge

Visc.: Brookfield RVT (#4, 20 rpm, 1 min) = 3090 cps
pH = 5.9.

Procedure:

Charge water and hexylene glycol into making tank and heat to 70-75 C. Dissolve citric acid and disperse **Lexate CRC**, stearyl stearate and **Lexol** with moderate agitation to avoid incorporating air. Cool to 45-50 C and add formalin and sodium chloride with good agitation. Continue stirring, cool to 40-45 C and add perfume and colors. Cool further with gentle agitation to 25-30 C. Pack into suitable containers and let stand. Viscosity develops fully after 24 h at 25 C.

Emollient Detergent Cream

Formula No. 1

(Nonionic-Anionic-Cationic)

A	**Veegum**	1.0
	Water	50.7
B	Cetyl Alcohol	0.3
	Stearyl Alcohol	0.3
	Lanacet	1.0
	Nimlesterol D	5.0
	Stearic Acid	4.4
	Cocoyl Sarcosine	3.3
	Pluronic F68	12.0
	Igepon AC-78	20.0
	Aromox C/12W	2.0
C	Preservative	q.s.

Procedure:
Add the **Veegum** to the water slowly, agitating continually until smooth. Heat A to 75 C. Heat B to 70 C, with slow mixing until uniform. Add A to B with slow agitation. Allow to cool to 40 C, and package while warm.

Directions for use:
Mix with warm water and rub into skin to lather. Rinse with water.

Formula No. 2

(Nonionic, W/O)

Oil Phase:	
Amerchol® L-500	2.0
Amerchol® C	5.0
Microcrystalline Wax (170-175 F m.p.)	10.0
Mineral Oil (180 visc.)	29.0
Sorbitan Sesquioleate	2.0
Water Phase:	
Water	50.0
Sorbitol Sol'n. (70%)	2.0
Perfume and Preservative	q.s.

Procedure:
Add the water phase at 85 C to the oil phase at 85 C while mixing. Continue mixing and cool to 40 C. Homogenize.

This soft, white w/o cream deposits an emollient film that provides protection against chapping and drying detergents. **Amerchol L-500** concentrated multisterol extract and **Amerchol C** multisterol absorption base add the moisturizing properties of lanolin alcohols and stabilize the w/o emulsion.

Emollient Clear Gel

(Nonionic)

Solulan C-24	15.0
Isopropyl Myristate	25.0
Oleyl Alcohol	5.0

G-1292	15.0
Propylene Glycol	5.0
Water	35.0
Perfume and Preservative	q.s.

Procedure:

Add the water at 50 C to the other ingredients at 50 C while mixing. After all water has been added, mix just until uniform.

Synthetic Liquid Hand Soap

(Nonionic-Anionic)

Water	88.00
Borax	0.40
ESI-Terge T-60	8.75
ESI-Terge B-15	2.75
Versene 100	0.10

Solids	8.4%
Active	8.4%
pH	8.0–8.6
Visc.:	25 cps max LV #1 Spindle 60 rpm @ 25 C

Procedure:

Add in order listed with adequate agitation, allowing each material to dissolve or disperse completely.

Waterless Hand Cleaner

Formula No. 1

(Nonionic)

Oil Phase:

Solulan 75	3.0
Glyceryl Monostearate (neut.)	5.0
Stearic Acid (XXX)	5.0
Cetyl Alcohol	2.0
Odorless Mineral Spirits	25.0

Water Phase:
Glycerin	5.0
Triethanolamine	2.0
Water	53.0

Perfume and Preservative	q.s.

Procedure:
Add the water phase at 85 C to the oil phase at 85 C while stirring. Continue mixing and cool to 32 C.

Note:
Add mineral spirits to oil phase after waxes have melted.

No. 2

(Nonionic)

A	Neo-Fat® 18-55	9.0
	Deodorized Kerosene	50.0
	Ethofat® 242/25	8.0
	Methyl Paraben	0.15
	Propyl Paraben	0.05
B	Perfume	q.s.
C	Sodium Hydroxide	0.65
	Water (deionized)	q.s. to 100.00

Procedure:
Heat A to 70 C. Heat C to 75 C. Add C to A with agitation and cool to 40 C. Add B mixing thoroughly.

No. 3

(Nonionic)

A	Mineral Oil (light)	20.00
	Deodorized Kerosene	20.00
	Stearic Acid (XXX)	5.00
	Lexemul 515	3.00
	Lexgard P	0.10

B Water   45.20
   Propylene Glycol   5.00
   Triethanolamine   1.50
   Lexgard M   0.20

Procedure:

Heat A to 60 C. Heat B to 65 C. Add A to B and blend until cool to 35 C. Fill.

No. 4

(Nonionic)

Oil Phase:
    Deodorized Kerosene   39.0
    Pale Oleic Acid   7.5
    Neodol® 25-3   2.0

Water Phase:
    Monoethanolamine   0.8
    Triethanolamine   2.6
    Propylene Glycol   2.5
    Glycerin   1.0
    Lanolin   0.5
    Butyl Oxitol® Glycol Ether   0.4
    Water   43.7

    Perfume and Color   as desired

Procedure:

Heat both oil and water phases separately to 70-75 C. Pour oil into water while stirring continuously. Continue until a homogeneous smooth gel is formed, pour into jars. Let cool to room temperature.

Stability:

Stable at 49 C/120 F for 7 days and after 5 days freeze-thaw.

No. 5

(Nonionic)

A Deodorized Kerosene   40.0
   Oleic Acid   9.0

Standamul® OXL (PPG-10-Ceteareth-20)	5.0
Lantox 55	1.0
Water	36.0
B Water	4.5
Triethanolamine (99%)	4.5
C Perfume Oil	q.s.
Dyes and Preservatives	q.s.

Procedure:

Heat A to 70 C under agitation. Heat B to 70 C under agitation. Add B to A and blend until uniform. Cool and at 40 C add individual components of C. Continue low speed agitation until product is uniform.

Note:

The propoxylated-ethoxylated alcohol provides cosolvency to the nonaqueous portion of this preparation. In addition, this fatty alcohol imparts dermal substantivity.

Pumpable Hand Cleaner Cream

(Nonionic)

Odorless Kerosene	20.00
Crodapur Lanolin	7.50
Polawax	5.00
Sodium Lauryl Ether Sulfate	4.00
Volpo N5	1.00
Water	62.50
Preservative and Perfume	q.s.

Procedure:

Heat oil phase to 65–70 C. Heat water phase to 70–75 C and add to oil phase with stirring. Avoid aeration and stir down to room temperature. Perfume at approximately 50 C.

Note:

Viscosity may be increased if required by the inclusion of ceto stearyl alcohol.

Aerosol Hand Cleaner

Crodolene LB 1120	6.00
Yeoman Lanolin	1.50
Deodorized Kerosene	41.00
Triethanolamine (special refined)	2.90
Propylene Glycol	2.80
Sodium Lauryl Sulfate	1.50
Water (deionized)	44.30
Perfume	q.s.

Procedure:

Heat oil phase to 65-70 C. Heat water phase to 65-70 C and add to oil phase under high speed mixer until cool.

Pack:

85% concentrate
15% Propellent 12/114 (50:50)

Liquid Hand Soap

Formula No. 1

Distilled Palm Kernel Fatty Acids	15.00
Tallow Fatty Acids	5.00
Crodolene LA 1010	2.50
Potassium Hydroxide	5.60
Empicol LQ27	10.00
Formalin	0.20
Water (deionized)	61.70
Perfume	q.s.

Procedure:

Blend all components except water, formalin and KOH by heating to 80 C. Heat water and KOH to similar temperature, then add to the molten oil phase. Stir until clear solution forms, then add formalin.

	No. 2	No. 3
		(Pearly)
Distilled Palm Kernel Fatty Acids	9.22	9.22
Tallow Fatty Acids	3.04	3.04

Crodolene LA 1010	1.52	1.52
Sodium Lauryl Sulfate	–	6.13
Monoethanolamine Lauryl Sulfate	6.13	–
Crodapearl	1.00	1.00
Morpholine	5.48	5.48
Water (deionized)	73.61	73.61
Formalin	q.s.	q.s.

Procedure:

Blend all components except morpholine by heating to 80 C with agitation. Add morpholine. Continue stirring until clear liquid forms, increasing percentage of morpholine if necessary. Perfume may be included at 45-50 C. Alternative amines may be employed to produce higher quality products. The pearl effect forms on standing for 4–12 h.

Opaque Lotion Soap

(Anionic)

A	**Lakeway 301-10**	25.00
	Pationic 138C	5.00
	Superamide 100 CG	4.00
	Pationic ISL	2.00
	Glycol Distearate	0.50
B	Water (deionized)	62.80
	Sodium Chloride	0.50
	Glydant 40-700	0.20
C	FD&C Blue No. 1	q.s.
	Perfume	q.s.

Visc.: (Brookfield Model RVT, Spindle #2,
@ 10 rpm @ 80 F   2000 cps

Procedure:

Combine ingredients of A, heat to ≈ 70 C to give a melt. Heat B to 70 C and add to B with agitation. Stir down to room temperature, add C and water loss.

Adjust pH with dilute sodium hydroxide to 7.0 if necessary.

Tar Remover Cream

White Oil (technical)	5.00
Kerosene (odorless)	20.00
Crodapur Lanolin	7.50
Polawax	10.00
Sodium Lauryl Sulfate	2.50
Paraffin Wax 140-145	2.50
Water (deionized)	52.50

Procedure:

Heat oil phase to 60 C. Heat water phase to 60 C. Add water to oils and continue stirring until emulsion thickens. Allow to cool and fill off.

Neutral pH Detergent Bars

Formula No. 1

(Anionic)

Igepon AC-78	25.0
Milled Bleached White Flour	52.5
Glycerin	3.0
Cornstarch	4.0
Lanolin (oil-soluble liquid fraction)	1.0
Isopropyl Myristate	2.0
Lactic Acid	2.0
Water	9.0

Note:

The pH of the bar is approximately 6.4.

No. 2

(Anionic)

Igepon AC-78	49.4
Sodium Dodecyl Benzene Sulfonate	4.87
Stearic Acid (XXX)	24.0
Water	4.4
Tallow Soap (anhydrous)	10.03
Miscellaneous (including sodium sulfate, perfume, and pigment)	7.0

Note:
The final pH of a 10% solution at 35 C is about 7.0.

Cosmetic Cleansing Bar

Formula No. 1

(Anionic)

A	Vancide 89RE	1.0
	Veegum F	1.0
B	Igepon AC-78 (83% solids)	57.3
C	Cetyl Alcohol	2.0
	Glyceryl Monostearate A.S.	5.5
	Stearyl Alcohol	7.5
	Modulan	3.0
	Polyethylene Glycol 6000	13.0
	Citric Acid	0.7
	Water	9.0

Procedure:
Blend A in part of B, then add the balance of B and blend well. Heat C to 70–75 C. Add A and B to C with agitation until uniform. Press into bar or cake.

Note:
The final pH should be about 5.0.

No. 2

(Anionic, Soap Base)

Soap Base 80/20 Tallow/Coco	95.9
Sopanox	0.1
Perfume Oil PA 53984	1.0
Pationic ISL	3.0
Polyox WSR N-80	1.5
Titanium Dioxide	0.5
pH:	10.1-10.3

Procedure:

Premix the **Pationic ISL** and perfume oil to reduce the viscosity of the **Pationic ISL** and facilitate subsequent incorporation. Combine the other materials to some degree of uniformity using suitable equipment. Add the **Pationic**-perfume mixture and continue blending until uniform. Extrude and stamp in the usual fashion.

No. 3

(Anionic, Economy Soap Base)

Soap Base 80/20 Tallow/Coco	80.00
Water	1.00
Pationic ISL	3.00
Nadex 360	12.00
Perfume PA 53984	1.00
Titanium Dioxide	0.50

Soap Bar

Formula No. 1

(Nonionic-Anionic)

Igepon AC-78	30
Igepal DM-970 Surfactant	5
Tallow/Coco Soap (90/10)	65

No. 2

Igepon AC-78	12
Igepal DM-970 Surfactant	20
Tallow/Coco Soap (90/10)	68

Note:

15-20% **Igepon AC-78** surfactant has been shown to give maximum lime soap dispersancy. The anionic should be added at the amalgamator. Total water content of about 12-13% gives good bar forming properties. The use of the coco soap may be eliminated due to the good lathering characteristics of the **Igepon** surfactant.

No. 3

(Nonionic-Anionic, Superfatted Neutral)

Tallow/Coco Soap (90/10)	10
Igepon AC-78	58
Igepal DM-970	2
Stearic Acid	30

No. 4

(Nonionic-Anionic, Superfatted Neutral)

Tallow/Coco Soap (90/10)	18
Igepon AC-78	42
Igepal DM-970	14
Stearic Acid	26

Note:

These bars can be made by adding the **Igepon AC-78** surfactant to the molten stearic acid/**Igepal** nonionic surfactant/soap mixture and pouring the resulstant slurry into molds. They could readily be made in a cooling plodder or cast in frames.

Syndet Bar

Formula No. 1

(Anionic)

Igepon AC-78	45.0
Nadex 360	22.0
Water (deionized)	12.0
Titanium Dioxide	0.5
Polyox WSR N-80	1.5
Pationic ISL	3.0
Perfume Oil PA 53984	1.0
Lathanol LAL	5.0
Patlac CA-95 NF	6.0
Lactic Acid (88%)	1.3
Sodium Lactate (60%)	1.0
pH	5.3-5.7

Procedure:

The wax portion of the formula is kept pliable after cooling by heating the cetyl alcohol and **Pationic ISL** together with stirring until uniform. Cool to 45–50 C. Add perfume. Hold at temperature until used.

The remainder of the formula is blended to some degree of uniformity. A ribbon blender, amalgamator, or sigma blender is recommended but a plodder can be used in the appropriate manner. The cetyl alcohol-**Pationic ISL** and the perfume mixture are then added and the final composition is worked until uniform.

The mixture may then be extruded and pressed in the usual fashion as though it were soap.

Note:

As with many specialty bar compositions there may be minor restrictions on bar configuration and the detail of the logo. Polyols or sodium lactate are recommended as die release agents.

No. 2

(Anionic, Lathanol)

		Mixing Order
Lathanol LAL-70 (powder)	24	1
Corn Starch or White Corn Dextrin	48	2
Nacconol 40DBX	5	3
Steol CS-460	5	7
Glycerin (cp)	9	6
Water	9	5
Perfume	as desired	8
Colorant	optional	4

Procedure:

Mix the dry ingredients (1, 2, 3, and 4) in a suitable mixer (e.g., a North-Master mixer). Prepare a solution of water, glycerin, **Steol CS-460**, and perfume. Add the solution to the dry mix. Mix until uniform. Transfer mixture to a soap plodder and extrude, using a moderate amount of heat on the plodder nose. Extruded bars can be cut to desired size. If desired, the extruded bar can be shaped by using very light pressure in a soap press.

No. 3

(Anionic)

		Mixing Order
Lathanol LAL-70 (powder)	20	2
Carbowax 6000	80	1
Perfume and Colorant	as desired	3

Procedure:

Break up **Carbowax 6000** (ca. 4 mesh). Add **Lathanol LAL-70** to **Carbowax 6000**. Mix on ball mill. Micropulverize the mixture, using a jump gap grid. Press into cakes of desired size and shape.

Framed Sea Water Soap

	Formula No. 1	No. 2
	(Anionic)	
Kettle Soap (33% water)	510	700
Nacconol 90F	270	280
Water	70	—
Sodium Sulfate	130	—
Sodium Chloride	20	20
Antioxidant (p-tert-butylphenol)	0.25	0.35
Color and Perfume	as desired	as desired

Procedure:
1. Weigh or pump kettle soap into crutcher.
2. Add **Nacconol 90F** gradually, withholding water until 1/2 to 2/3 of **Nacconol 90F** has been added.
3. Add water and balance of **Nacconol 90F**.
4. Continue crutching until a smooth pasty soap mixture is obtained.
5. Add dry salts, crutch until well mixed.
6. Drop into frames.
7. When hard, cut into bars.

(In preparing Formula No. 2, Step 2, add all the **Nacconol 90F**, proceeding to Step 4.)

Milled Sea Water Soap

	Formula	No. 1	No. 2
Toilet Soap Stock (10-12% water)		60	70
Nacconol 90F		27	30
Sodium Sulfate		13	–
Color and Perfume		as desired	as desired

Procedure:

Mix **Nacconol 90F** and sodium sulfate with toilet soap base in amalgamator or other suitable mixer. Mill until uniform on a soap mill provided with adequate power. Plot into bars of desired shape and weight in a standard soap plodder.

Shaving Cream

Formula No. 1

(Anionic, Lather)

A Water	47.5
Sodium Hydroxide (19% aqueous sol'n.)	7.0
Stearic Acid (USP)	6.5
Glycerin (USP)	2.0
Isopropyl Myristate	2.0
Standapol® SHC-101	35.0
B Perfume Oil	q.s.
Dyes and Preservatives	q.s.

Procedure:

Heat water in A to 80 C. Keep temperature constant. Add remaining ingredients of A, one at a time, under agitation. Cool, continue agitation, and adjust pH to 9.0 ± 0.5 with 50% citric acid aqueous solution. At 45-50 C add individual components of B under agitation. Continue low-speed agitation until product is uniform.

Note:

The sulfosuccinate-anionic blend inhibits dermal irritation while also providing synergistic detergency to this system.

		No. 2	No. 3	No. 4
		(Nonionic, Aerosol Foam)		
A	**Hydrofol** Acid 1895			
	(95% stearic)	5.73	5.40	4.20
	Hydrofol Acid 1495			
	(95% myristic)	1.60	1.20	–
	Hydrofol Acid 1295			
	(95% lauric)	–	–	0.40
	Starfol EGMS	–	–	0.80
	Starfol IPM	–	–	1.00
B	Triethanolamine	–	3.80	2.40
	Aminomethyl Propanediol			
	(AMPD)	2.95	–	–
	Polyglycol 15-200	6.00	6.00	–
	Sorbitol (70%) USP	–	–	3.00
	Propylene Glycol USP	–	–	3.00
	Water (demineralized)	83.70	83.60	85.20
C	Perfume	q.s.	q.s.	q.s.

Procedure:

Dissolve all material in Part B in the water and heat to 72-75 C. Melt all components in Part A together at 75-78 C. Add A to B slowly with moderate and smooth agitation. Allow to cool with gentle mixing. Add perfume at 40-45 C. Put concentrate into aerosol bottle or appropriately lined can and crimp on the valve. Pressure fill the propellant blend through the valve.

Concentrate	92.0% (wt.)
Propellant 12/114 (60:40)	8.0%

No. 5

(Nonionic, Women's)

A	**Veegum**	3.0
	Water	83.5
B	Glycerin	2.0

	Sorbitol (70% sol'n.)	3.0
	Triton X-100	3.0
C	**Myrj 45**	5.0
D	**Vancide 89RE**	0.5

Procedure:

Add the **Veegum** to the water slowly, agitating continually until smooth. Add B to A and heat to 70 C. Heat C to 75 C, add to A and B and mix. Cool with agitation and add D at approximately 50 C. Continue mixing until cool.

No. 6

(Nonionic-Anionic, Aerosol)

A	**Darchem 12**	7.16
	Glycerin	2.71
	Emery 622	1.00
	Pationic ISL	2.00
	Water (deionized)	79.81
B	Potassium Hydroxide (34.2%)	4.61
	Sodium Hydroxide (19.1%)	0.96
C	**Clindrol 100C**	1.00
	Coconut Oil (76°)	0.25
	Perfume F77-155	0.50

Procedure:

Combine Part A, holding out approximately 25% of water and heat to 75 C with moderate stirring. Combine Part B, heat to 75 C and add to Part A with agitation. Rinse vessel with about half of water held back from Part A. Combine Part C, warm to 45 C and add to mixture A/B at 45 C. Rinse vessel with remaining water. Stir down to room temperature.

Important:

Titrate and adjust to 0.1 to 0.2% free fatty acid as KOH.

For packaging:

Soap concentrate	97%
Propellant, isobutane/propane (87:13 ratio)	3%

No. 7

(Nonionic-Cationic, Brushless)

A	Stearic Acid (XXX)	8.00
	Lexate IL	4.00
	Lexemul 561	3.00
	Stearyl Alcohol	1.00
B	Glycerin	10.00
	Triethanolamine (85%)	0.75
	Lexgard M	0.15
	Lexgard P	0.05
	Water	up to 73.05
	Perfume	q.s.

Consistency: very soft, pearlescent, short fiber cream.

Procedure:

Weigh and melt the ingredients of A, stir until homogeneous and heat mixture to 70–75 C. Charge the ingredients of B into a separate vessel equipped with an agitator and provisions for heating and cooling. Dissolve the **Lexgards** with heating and agitation and bring temperature of completed B to 70–75 C. Gradually add A to B with vigorous agitation and when addition is complete, reduce agitation and cool to 40–45 C. Add and disperse perfume as required; cool to 35 C and package. Consistency develops fully after 24 h at room temperature.

Emulsifier-Free Shaving Gel

(Nonionic)

A	**Veegum**	5.0
	Water	82.0
B	**Amerchol L-101**	2.5
	Acetulan	0.5
	Silicone (350 cs)	10.0
	Preservative	q.s.

Procedure:

Add the **Veegum** to the water slowly, agitating continually until smooth. Continue stirring, add B and mix until uniform.

Directions for use:
Wash face thoroughly with soap and water. Leave face wet and apply gel sparingly.

Shaving Soap

(Anionic)

A	Neo-Fat® 65	28.7
	Neo-Fat® 255	6.1
	Neo-Fat" 18-43	10.0
	KOH	9.75
	BHA	0.1
	Methyl Paraben	0.2
B	Perfume	q.s.
C	Siponate SGS	5.0
	Versene	0.2
	Water (deionized)	q.s. to 100.0

Procedure:
Heat A to 70 C. Heat C to 75 C. Add C to A with agitation and cool to 40 C. Add B mixing thoroughly.

Skin Cleanser

(Nonionic–Anionic–Amphoteric)

A	Water	48.5
	Standapol® 130-E	10.0
	Standapol® ES-2	20.0
	Standapol® BAW	10.0
	Standamid® SD	3.5
	Standamox® CAW	2.0
	Polyquart® H-7102	5.0
	Ethylene Glycol Monostearate	1.0
B	Perfume Oil	q.s.
	Dyes and Preservatives	q.s.

Procedure:
Heat water to 75-80 C. Keep temperature constant. Add other ingredients of A in above listed order of addition, under constant agitation.
Cool, continue stirring, and at 45 C add individual components of B. Adjust the pH to 6.0 ± 0.5 with 50% citric acid aqueous solution and continue
low agitation until product is uniform.

Note:
Dermal and ocular irritation data available upon request. This blend of
ethoxylated sulfates, betaine, and amide with amine oxide provides synergistic detergency and mildness. Addition of the polyamine quaternary
blend provides dermal substantivity.

Skin Cleanser and Moisturizer

Formula No. 1

(Amphoteric)

Water	37.0
Standamul® O-20	10.0
Standapol® OLB-50	40.0
Standamul® O5	5.0
Propylene Glycol	5.0
Hygroplex HHG®	3.0
Perfume Oil	q.s.
Dyes and Preservatives	q.s.

Procedure:
Blend ingredients in order listed at 45-50 C. Continue stirring until
product reaches room temperature.

Note:
The blend of betaine, ethoxylated oleyl alcohol emulsifiers, and CLR
moisturizer, provides efficient detergency and mildness to this clear viscous
liquid cleanser.

No. 2

(Amphoteric)

Water	37.0
Standamul® O-20	10.0

Standapol® OLB-50	40.0
Standamul® O5	5.0
Propylene Glycol	5.0
Hygroplex HHG®	3.0
Perfume Oil	q.s.
Dyes and Preservatives	q.s.

Procedure:

Blend ingredients in order listed at 45-50 C. Continue stirring until product reaches room temperature.

Note:

The blend of betaine, ethoxylated oleyl alcohol emulsifiers, and CLR moisturizer, provides efficient detergency and mildness to this clear viscous liquid cleanser.

Emollient Body Cleanser

(Nonionic-Anionic-Amphoteric)

Water	62.0
Sodium Chloride	3.0
Standapol® BAW	20.0
Standapol® WAS-100	10.0
Standamul® HE	3.0
Standamid® LD (premelted at 45 C)	2.0
Perfume Oil	q.s.
Dyes and Preservatives	q.s.

Procedure:

The order of addition is given above. Add all materials singly under adequate agitation. Adjust the pH to 7.0 ± 0.5 with 50% citric acid aqueous solution. Continue agitation until product is homogeneous.

Note:

The betaine inhibits dermal irritation while also providing synergistic detergency to this combination of amide and anionic. The ethoxylated cocoate provides emollient and substantive effects when combined with the betaine.

Pearlescent Body Cleanser

(Nonionic-Anionic)

A	Water	24.0
	Standapol® ES-2	20.0
	Standamid® SD	4.0
	Standapol® CS	50.0
	Ethylene Glycol Monostearate	2.0
B	Perfume Oil	q.s.
	Dyes and Preservatives	q.s.

Procedure:

Blend Part A thoroughly at 80 C using the given order of addition. Continue blending while product cools slowly to 40 C. AT 40 C add individual components of Part B under agitation. Continue sweep-type agitation until product reaches room temperature.

Note:

The anionic sulfate-EGMS blend provides pearlescence and emollient dermal effects.

Sudsing Cleansing Lotion

(Anionic)

Stearic Acid	13.0
Triethanolamine	3.0
Water	67.5
Glycerin	6.5
Igepon® AM-78	10.0

Procedure:

Add stearic acid and triethanolamine to water at 40–50 C. Stir to solution. Add glycerin and **Igepon** surfactant. Cool with stirring to room temperature.

Note:

A colloid mill or homogenizer should be used.

Acidic Skin Cleansing Lotion

(Nonionic-Anionic)

A	Water	46.0
	Sodium Chloride	1.0
	Standapol® SHC-101	50.0
	Standamid® LD	2.0
	Ethylene Glycol Monostearate	1.0
B	Perfume Oil	q.s.
	Dyes and Preservatives	q.s.

Procedure:

Blend the ingredients of Part A, one at a time, at 80 C. Cool, continue stirring, and at 40 C add individual components of Part B. Blend until uniform and adjust the pH to 5.5 ± 1.0 with 50% citric acid aqueous solution. Continue low speed agitation until product reaches room temperature.

Note:

The sulfosuccinate-anionic base provides efficient detergency and mildness for this pearly liquid foaming cleanser.

Cleansing Lotion

(Nonionic)

A	**Crodafos CAP**	1.0
	Kesscolin™ Absorption Base	15.0
	Petrolatum (USP white)	15.0
	Propyl Paraben	0.1
B	Methyl Paraben	0.1
	Water (deionized)	q.s. to 100.0
C	**Dowicil 200**	0.1
	Water (deionized)	1.0
D	Perfume	q.s.

Procedure:

Heat A to 70 C. Heat B to 75 C. Add B to A with agitation. Cool to 50 C. Add C, cool to 40 C. Add D.

Cleansing Cream

	Formula No. 1	No. 2
		(Nonionic, Enriched)
Oil Phase:		
Amerchol L-101	3.0	3.0
Amerlate P	1.0	1.0
Beeswax (USP)	10.0	17.0
Mineral Oil (70 visc.)	43.15	21.5
Isopropyl Palmitate	–	21.5
Ozokerite	7.0	–
Glyceryl Monostearate (s.e.)	2.0	–
Glyceryl Monostearate (neut.)	–	2.0
Water Phase:		
Triethanolamine	0.25	–
Borax (USP)	0.6	1.0
Water	33.0	33.0
Perfume and Preservative	q.s.	q.s.

Procedure:

Add the water phase at 75 C to the oil phase at 75 C while stirring. Continue mixing and cool to 30 C.

Firm but light-textured glossy enriched creams for cleansing and moisturizing. Unusual slip for ease of application.

No. 3

(Nonionic)

A **Neo-Fat®️ 18-55**	3.0
Kesscolin™ Absorption Base	3.0
Stearyl Alcohol	3.0
Carnation Mineral Oil	15.0
Propyl Paraben	0.1
B Methyl Paraben	0.1
Carbopol 934	0.5
Glycerin	5.0
Water (deionized)	q.s. to 100.0

C Triethanolamine	1.5
D **Dowicil 200**	0.1
Water (deionized)	1.0
E Perfume	q.s.

Procedure:
Heat A to 70 C. Heat water to 75 C. Add glycerin and methyl paraben. Disperse **Carbopol.** Add B to A with agitation. Add C, cool to 50 C. Add D, cool to 40 C. Add E.

No. 4

(Nonionic, Washable)

Oil Phase:

Solulan® 75	2.0
Sonojell #9	7.5
Tween 60	1.0
Cetyl Alcohol	7.0
Cerasynt SE	7.0
Isopropyl Myristate	7.5

Water Phase:

Glucam® E-20	2.5
Water	63.0
Propylene Glycol	2.5
Perfume and Preservative	q.s.

Procedure:
Heat both phases to 75 C. Add water phase to oil phase while stirring. Allow to cool with continued agitation to 45 C. Add perfume, stir to 38 C.

No. 5

(Nonionic)

A Beeswax	10.0
Kessco® X-675	1.0
Mineral Oil 70/80	40.0

Kesscolin™	2.0
Methyl Paraben	0.1
Propyl Paraben	0.1
B Polysorbate 60	0.25
Borax	1.0
Water (deionized)	q.s. to 100.0

Procedure:
Melt and blend A to 60 C. Combine B at 60 C and blend A into B. Mix until cool.

No. 6

(Nonionic)

A Beeswax	15.0
Petrolatum	15.0
Mineral Oil	15.0
Kessco® Isopropyl Myristate	3.0
Kessco X-653 or X-654	1.5
Kesscolin™	1.5
Methyl Paraben	0.1
Propyl Paraben	0.1
B Borax	1.25
Polysorbate 80	0.3
Glycerin	3.0
Water (deionized)	q.s. to 100.0

Procedure:
Blend A at 70 C. Heat water and B to 70 C. Add water mixture to A with agitation. Continue to mix until cool 35-40 C.

No. 7

(Nonionic, Liquifying)

Kessco® Diethyleneglycol Monostearate	2.0
Kessco PEG 400 Monostearate	2.0
Kessco Butyl Stearate Cosmetic	5.0

Kessco Isopropyl Myristate	45.0
Kesscolin^{TM}	6.0
White Petrolatum	20.0
Paraffin	20.0
Perfume	q.s.

Procedure:
Blend ingredients and melt at 80 C adding perfume when mixture has cooled.

No. 8

(Nonionic-Anionic)

A	**Neo-Fat® 18-55**	14.4
	Kessco® Isopropyl Myristate	5.0
	Kesscolin^{TM}	4.0
	Methyl Paraben	0.1
	Propyl Paraben	0.5
B	Perfume	q.s.
C	Propylene Glycol	5.0
	Veegum	0.25
	Triethanolamine	3.0
	Water (deionized)	q.s. to 100.0

Procedure:
Heat A to 70 C. Heat C to 75 C. Add C to A with agitation and cool to 40 C. Add B mixing thoroughly.

No. 9

(Nonionic-Anionic)

A	**Kesscolin**^{TM}	5.0
	Beeswax (USP)	12.0
	Petrolatum	15.0
	Kessco® Isopropyl Myristate	5.0

Kessco Butyl Stearate Cosmetic	2.0
Mineral Oil (light)	23.0
Kessco Glyceryl Monostearate Pure	1.7
Propyl Parahydroxybenzoate	0.05
Methyl Parahydroxybenzoate	0.1

B Perfume — q.s.

C Maprofix WAC	1.0
Borax	0.6
Triethanolamine	0.5
Water (deionized)	q.s. to 100.0

Procedure:

Heat A to 70 C. Heat C to 75 C. Add C to A with agitation and cool to 40 C. Add B mixing thoroughly.

No. 10

(Nonionic-Cationic)

A Glyceryl Monostearate	8.0
Emulgin® B-1	4.0
Standamul® G	10.0
Mineral Oil (NF)	20.0

| B Water | 53.0 |
| Polyquart® H-7102 | 5.0 |

| C Perfume Oil | q.s. |
| Dyes and Preservatives | q.s. |

Procedure:

Heat Part A to 75-80 C under agitation. Heat Part B to 75-80 C under agitation. Add Part A to Part B, blend until uniform. Cool, and at 45 C add individual components of B under agitation. Continue low speed agitation until product reaches room temperature.

Note:

This o/w cream provides a nongreasy application. The polyamine quaternary blend in addition, imparts dermal substantivity and conditioning effects. The emulsifier-branch chain alcohol blend also yields efficient cleansing properties to this system.

No. 11

(Nonionic-Cationic)

A	Dehydag® Wax SX	10.0
	Standamul® G	10.0
	Mineral Oil (NF)	15.0
	White Petrolatum (USP)	2.0
B	Water	63.0
C	Perfume Oil	q.s.
	Dyes and Preservatives	q.s.

Procedure:

Heat Part A in a suitable vessel under constant agitation to 60 C. Heat Part B in a separate vessel to 60 C and slowly add to Part A, under moderate agitation. Continue agitation with addition of C at 40 C. Blend until uniform.

Note:

This o/w cream provides a nongreasy application. The emulsifier-branch chain alcohol blend also imparts efficient cleansing properties to this system.

No. 12

(Nonionic-Cationic)

A	Glyceryl Monostearate	8.0
	Emulgin® B-1	4.0
	Standamul® G	10.0
	Mineral Oil (NF)	20.0
B	Water	53.0
	Polyquart® H-7102	5.0
C	Perfume Oil	q.s.
	Dyes and Preservatives	q.s.

Procedure:

Heat Part A to 75–80 C under agitation. Heat Part B to 75–80 C under agitation. Add A to B, blend until uniform. Cool, and at 45 C add individual components of Part B under agitation. Continue low speed agitation until product reaches room temperature.

Note:
This o/w cream provides a nongreasy application. The polyamine quaternary blend in addition, imparts dermal substantivity and conditioning effects. The emulsifier-branch chain alcohol blend also yields efficient cleansing properties to this system.

No. 13

(Anionic)

A	Mineral Oil (70/80 cps visc.)	47.1
	Kesscowax™ B or Natural Beeswax	9.3
	Paraffin (122–127 F M.P.)	6.0
B	Carbopol 934	0.2
C	Borax	1.7
	Water	q.s.

Procedure:
Melt A at minimum temperature for melting waxes. Disperse B with vigorous agitation into water at 70 C. Dissolve C in small amount of water. Add B to A with agitation to form emulsion, then add C with stirring.

No. 14

(Nonionic-Anionic)

A	Kesscolin™ Absorption Base	2.0
	Myrj 52-S	3.0
	Carnation Mineral Oil	40.0
	Propyl Paraben	0.1
B	Glycerin	5.0
	Carbopol 934	0.6
	Methyl Paraben	0.1
	Water (deionized)	q.s. to 100.0

C Triethanolamine	0.7
D Perfume	q.s.

Procedure:

Heat A to 70 C. Heat water to 75 C. Add methyl paraben and disperse **Carbopol**. Add B to A with agitation, followed by C. Cool with continued agitation to 40 C. Add D.

No. 15

(Nonionic-Anionic, Washable)

Oil Phase:

Amerlate® P	3.0
Standamul 1414E	5.0
Isopropyl Myristate	8.0
Cetyl Alcohol	8,0
Ceraphyl 140-a	3.0
Cerasynt SD	4.5

Water Phase:

Glucam® E-20	5.0
Water	58.5
Propylene Glycol	2.0
Standapol SHC 101	3.0

Perfume and Preservative	q.s.

Procedure:

Heat both phases to 75 C. Add water phase to oil phase while stirring. Allow to cool with continued agitation to 45 C. Add fragrance, stir to 38 C.

Biostatic Skin Cleanser

(Nonionic-Anionic)

A **Veegum K**	1
Water	22
Triton X-202 (30% solids)	66
Propylene Glycol	5

B	Vancide 89RE	2
	Amerchol L-101	1
	Solulan 16	3

Procedure:
Add the **Veegum K** to the water slowly, agitating continually until smooth. Add the rest of A and heat to 55 C. Disperse the **Vancide 89RE** in the B phase while heating to 60 C. Add B to A and stir until cool.

Oily Skin and Pore Cleaner

(Nonionic)

A	Polyethylene Powder	5.0
	Polysorbate 80	3.0
	Kessco® Isopropyl Myristate	2.0
	Maprofix WAC	0.25
	Titanium Dioxide	1.5
	Propylene Glycol	7.5
	Methyl p-Hydroxybenzoate	0.1
	Propyl p-Hydroxybenzoate	0.05
B	Perfume	q.s.
C	**Veegum**	5.0
	Water (deionized)	q.s. to 100.0

Procedure:
Heat A to 70 C. Heat C to 75 C. Add C to A with agitation and cool to 40 C. Add B mixing thoroughly.

Facial Cleanser

Formula No. 1

(Anionic)

A	**Standapol® ES-2**	54.0
	Carbowax 400	10.0
	Standamid® KD	2.0
	Water	6.0

B Ethylene Glycol Monostearate	5.0
Dehydag® Wax SX	10.0
Mineral Oil (NF)	10.0
C Placenta Liquid Water Soluble®	3.0
D Perfume Oil	q.s.
Dyes and Preservatives	q.s.

Procedure:

Heat Part A to 75 C under agitation. Heat Part B to 75 C under agitation. Add B to A, blend and maintain temperature for 15 min. Cool, continue blending, and at 45 C add C. Continue low speed agitation and at 40 C add individual components of Part D. Blend until product is uniform.

Note:

This o/w cream provides a nongreasy application. The emulsifier-anionic blend yields excellent cleansing properties to this system. The CLR Placenta Liquid® imparts unique dermal effects.

No. 2

(Nonionic)

A Kessco® PEG 400 Monostearate	3.6
Mineral Oil (light)	1.0
Kessco Glycerol Monostearate (s.e.)	10.0
Propyl Paraben	0.1
Methyl Paraben	0.1
B Perfume	q.s.
C Propylene Glycol	6.0
Water (deionized)	q.s. to 100.0

Procedure:

Heat A to 70 C. Heat C to 75 C. Add C to A with agitation and cool to 40 C. Add B mixing thoroughly.

No. 3

(Nonionic)

| A Neo-Fat® 18-55 | 12.0 |

Kessco® X-653 or X-654	4.0
Mineral Oil	15.0
Methyl Paraben	0.1
Propyl Paraben	0.1
B Sodium Lauryl Sulfate (100%)	2.0
Triethanolamine	1.5
Glycerin	5.0
Water (deionized)	q.s. to 100.0

Procedure:
Blend A to 70 C. Heat water and B to 70 C. Add water mixture to A with agitation. Continue to mix until cool 35-40 C.

No. 4

(Nonionic-Anionic)

A Pationic 138C	6.0
Hamposyl 2-30	25.0
Emerest 2355	3.0
Pationic ISL	3.0
Clindrol Superamide 100 CG	3.0
Mapeg 6000 DS	1.0
B Methyl Paraben	0.2
Water (deionized)	58.6
C Perfume #03978	0.2
Lactic Acid (44%)	q.s.

Visc.: (Brookfield Model RVT Spindle #2 @ 5 rpm @ 80 F)	7000 cps

Procedure:
Combine ingredients of A and heat to 70 C. Combine ingredients of B and heat to 72 C. Add B to A with good agitation. Add perfume at 45 C and continue stirring to room temperature. Adjust pH to 5.6-5.8 with lactic acid.

No. 5

(Nonionic-Cationic)

A Oil Phase:	
White Petrolatum	7.00
Adol 52-NF	7.00
Varonic 1000 MS (23 POE stearic acid)	6.00
B Water Phase:	
Glycerol	8.00
Methyl Paraben (USP)	0.10
Adogen 432-CG	0.05
Water (demineralized)	to 100.00
C Perfume	q.s.

Procedure:

Heat both phases to 75 C. Add oil phase to water phase slowly with agitation. Mix 5 min. Cool with agitation to 40 C. Add perfume and pour into jars.

No. 6

(Nonionic-Cationic)

A Mineral Oil 125/135	15.00
Lexemul 561	10.00
Lexate PX	5.00
Lexate IL	5.00
B Glycerin	5.00
Lexgard M	0.15
Lexgard P	0.05
Water	up to 59.80
Perfume	q.s.

Visc.: Brookfield RVT (#3, 10 rpm, 1 min) = 3900 cps
pH = 5.4.

Procedure:

Weigh and melt the ingredients of A, stir until homogeneous and heat mixture to 70-75 C. Charge the ingredients of B into a separate vessel equipped with an agitator and provisions for heating and cooling. Dissolve the **Lexgards** with heating and agitation and bring temperature of completed B to 70-75 C. Gradually add A to B with vigorous agitation and when addition is complete, reduce agitation and cool to 30-35 C. Add and disperse perfume as required, cool to room temperature and package. Consistency develops fully on standing 24 h.

No. 7

(Nonionic-Cationic)

A	Mineral Oil (light)	20.00
	Lexemul 561	12.00
	Lexol PG8-10	5.00
	Cetyl Alcohol	3.00
	Pluronic L64	1.00
B	Glycerin	6.00
	Lexgard M	0.15
	Lexgard P	0.05
	Water, Perfume	q.s. to 100.00

Procedure:

Weigh and melt together the ingredients of A, stir until homogeneous and heat mixture to 65-70 C. Charge the ingredients of B into a separate vessel equipped with an agitator and provisions for heating and cooling. Dissolve the **Lexgards** with heating and agitation and bring temperature of completed B to 65-70 C. Gradually add A to B with vigorous agitation, and when addition is complete, reduce agitation and cool to 40-45 C. Add and disperse perfume as required; cool to 30-35 C and package. Consistency develops fully after standing 24 h at room temperature.

Note:

Consistency: Very soft, glossy, medium-short fiber cream.

No. 8

(Nonionic-Anionic-Amphoteric)

A	**Lexol PG 8-10**	2.00
	Mineral Oil (light)	2.00
	Lexgard P	0.05

Lexemul P	3.50
Stearic Acid (XXX)	13.40
B **Lexgard M**	0.15
Bronopol	0.05
Lexein P50	2.50
Glycerin	1.70
Triethanolamine	0.75
Perfume	0.25
Water	73.65

Procedure:

Prepare Parts A and B separately, withholding 25% of the water, **Lexein P50**, **Bronopol**, and perfume. Heat both parts to 70-75 C and mix until homogeneous. Add Part A to Part B with vigorous agitation and cool to 50 C. Take the remaining 25% of the water and **Lexein P50** and warm with mixing until dissolved. Add this latter solution to batch at 50 C. Cool batch to 40 C and add perfume with stirring. Continue stirring, cool batch to 30 C and fill.

	No. 9 *(Anionic-Amphoteric, Clear Viscous Liquid)*	No. 10 *(Anionic-Amphoteric, Clear Gel)*
Water	50.3	52.3
Monateric CSH-32	33.3	16.7
Monamate CPA-40	5.0	—
Monamate OPA-30	—	16.7
Monateric ISA-35	11.4	14.3
Approx. Visc. @ pH 6.0	5000 cps	35,000 cps

Procedure:

Add ingredients in the order listed with agitation, and heat to 50 C before adding **Monateric ISA-35**. Adjust to desired pH with phosphoric acid (the gel should be adjusted and packed while warm and fluid).

No. 11

(Nonionic-Amphoteric, Protein)

Lexmul 503	10.00

Cetyl Alcohol	5.00
Brij 58	5.00
Mineral Oil (light)	8.25
Pluronic L-42	3.00
Lexgard M	0.15
Lexgard P	0.05
Lexein X250	5.00
DC 200 Fluid	0.20
Perfume G73-147[1]	0.25
FD&C Yellow No. 5	q.s.
Water	q.s. to 100.00

[1] Perry Bros.

Procedure:
Heat water to 65–70 C, add all ingredients with agitation except **Lexein X250,** perfume and dye. Cool batch with agitation to 45 C and add balance of ingredients. Cool batch with agitation to 30 C and fill.

Note:
Visc.: If a lower viscosity is desired up to 10% ethyl alcohol can be incorporated into this formula.

Facial Freshener and Toner

(Nonionic–Cationic, Two-Layer)

A Top Layer:	
Standamul® G	80.00
Standamul® 318	17.00
Perfume Oil	3.00

B Bottom Layer:	
Water	83.99
Ethyl Alcohol–SD 40	16.00
Citric Acid (50% aqueous sol'n.)	0.01

Procedure:
Blend A. Blend B. Add 6.4 parts of Part A (top layer) to 93.6 parts of Part B (bottom layer) under agitation.

Note:

The branch chain alcohol provides both emolliency and unique solubilization properties in this two layer freshener. The fatty acid triglyceride imparts emollient dermal effects.

Facial Washing Cream and Make-Up Remover

(Anionic)

A	Water	57.6
	Propylene Glycol	7.0
	Standapol® SHC-101	3.0
	Standamul® 1414-E	10.0
	Mineral Oil (NF)	14.0
	Cetyl Alcohol	4.0
	Stearyl Alcohol	4.0
	Silicone 200 Fluid (350 cs)	0.4
B	Perfume Oil	q.s.
	Preservatives	q.s.

Procedure:

Blend A at 75 C in the order given, under constant agitation. Cool and at 50 C add individual components of B under agitation. Continue sweep-type agitation until product cools to room temperature.

Note:

This o/w cream provides emolliency with the blend of fatty alcohols and ethoxylated myristate. The sulfosuccinate-anionic blend also imparts mild and efficient detergency.

Make-Up Remover–Hydrocarbon Type Gel

(Nonionic)

A	Mineral Oil 70/80	30.0
	Kessco® X-675	10.0
	Perfume	q.s.
	Kesscolin™	10.0
B	Petrolatum	q.s. to 100.0

Procedure:
Heat all ingredients at 60 C. Stir until cool.

Liquid Herbal Make-Up Remover

Standamul® G	40.0
Mineral Oil (NF)	57.0
Calendula Oil CLR®	3.0
Perfume Oil	q.s.
Preservatives	q.s.

Procedure:
Blend ingredients in the order given, under adequate agitation. Continue stirring until product is homogeneous.

Note:
The branch chain alcohol provides emolliency in addition to its solvent properties.

Liquid Eye Shadow Remover

Standamul® G	40.0
Mineral Oil (NF)	60.0
Perfume Oil	q.s.
Preservatives	q.s.

Procedure:
Blend ingredients in the order given, under adequate agitation. Continue stirring until product is homogeneous.

Note:
The branch chain alcohol provides emolliency in addition to its solvent properties.

Eye Make-Up Remover

(Amphoteric)

Miranol® 2MHT Modified	5.0-10.0
Tween 20	1.0- 2.0
Propylene Glycol	1.0- 2.0

Mixed Parabens	0.1- 0.1
Ethylenediamine Tetra-Acetic Acid	0.1- 0.1
Water	92.8-85.8

pH: 7-8

Face Mask

(Nonionic)

Liquid Phase:	
Solulan 98	5.0
Ethyl Alcohol	5.0
Menthol	0.1
Water	63.4
Glycerin	2.5
Methocel 60HG (400 cps)	1.5
Powder Phase:	
Bentonite 660	15.0
Zinc Oxide (USP)	5.0
Titanium Dioxide 3328	2.5
Perfume and Preservative	q.s.

Procedure:

Disperse the **Methocel** in the glycerin. Add the **Solulan 98** to the alcohol and menthol, stir to dissolve, then add the water. Mix well. Add this solution to the **Methocel** dispersion, then add this mixture in increments to the combined powders, working well after each addition.

Note:

A soothing soft paste with good performance.

Acne Scrub Cream

(Nonionic-Anionic)

A	**Veegum**	2.0
	Water	58.2
B	Propylene Glycol	10.0
	Amerchol L-101	15.0

Aldo MSA	3.0
Triton X-202	1.4
C AC 9A Polyethylene	10.0
D Eucalpytus Oil	0.4
Preservative	q.s.

Procedure:
Prepare A by adding the **Veegum** slowly to the water, agitating continually until smooth. Heat to 75 C. Heat B ingredients to 70 C. Add B to A with mixing until smooth. Add C and D with mixing until cool.

Toothpaste

Formula No. 1

(Anionic)

A Water	25.30
CMC 12 HP	1.25
Sorbitol Solution (USP)	23.75
Dicalcium Phosphate Dihydrate	48.00
Preservatives	q.s.
B **Texapon® K-12**	1.70
Flavor, Sweetening Agent	q.s.

Procedure:
Heat water to 80-85 C. Sprinkle in CMC and hydrate completely. Continue stirring, add sorbitol and preservatives, blend until uniform. Sprinkle in dicalcium phosphate dihydrate very slowly. Mix until smooth under low agitation. Cool at 40-45 C. De-aerate under vacuum. Add individual components of B under low agitation. Blend until uniform. Fill.

Note:
The SLS powder aids in removal of oral debris in this dentifrice base because of its excellent foaming properties.

No. 2

A **Veegum**	1.0
Water	18.5

B	Sodium Carboxymethylcellulose (high visc.)	0.5
	Glycerin	30.0
C	Dicalcium Phosphate	47.0
D	Flavor	1.0
E	Sodium Lauryl Sulfate	2.0
	Preservative	q.s.

Procedure:

Add the **Veegum** to the water slowly, agitating continually until smooth. Wet the CMC with glycerin and add to A slowly with agitation. Add C slowly with agitation. Add D and mix. Add E slowly with just enough mixing to obtain a smooth batch.

No. 3

A	**Veegum F**	1.20
	Sodium Carboxymethylcellulose (med. visc.)	0.70
	Water	23.40
B	Saccharin	0.15
	Water	2.00
C	Sorbitol (70% sol'n.)	12.50
	Glycerin	12.50
D	Dicalcium Phosphate Dihydrate	45.00
E	Flavor	1.00
F	Sodium Lauryl Sulfate	1.50
	Preservative	q.s.

Procedure:

Dry blend the **Veegum F** and the CMC. Add to the water slowly, agitating continually until smooth. Add B to A. Add C and D alternately to A and B with mixing. Add E to the mixture, mixing until smooth. Add F to the other components with a minimum of slow mixing to avoid incorporation of air.

Note:

This formula will have a thin mixing viscosity but, due to the thixotropic nature of **Veegum**, will set up in the tube.

No. 4

(Low Foam)

A	**Veegum**	1.50
	Sodium Carboxymethylcellulose (med visc.)	0.80
	Water	28.05
B	Saccharin	0.15
	Water	2.00
C	Sorbitol (70% sol'n.)	15.00
	Glycerin	10.00
D	Dicalcium Phosphate Dihydrate	30.00
	Tricalcium Phosphate	10.00
E	Flavor	1.00
F	Sodium Lauryl Sulfate	1.50
G	preservative	q.s.

Procedure:

Dry blend the **Veegum** and the CMC. Add to the water slowly, agitating continually until smooth. Add B to A. Add C and D alternately to A and B with mixing. Add E to the mixture and mix until smooth. Add F to the other components with a minimum of slow mixing. Add G.

Note:

This formula is for use with electric toothbrushes.

No. 5

(Aerosol)

A	**Veegum**	2.0
	Water	19.3
B	Sodium Carboxymethylcellulose (high visc.)	0.2

Sorbitol (70% sol'n.)	15.0
Glycerin	15.0
C Dicalcium Phosphate Dihydrate	45.0
D Flavor	1.0
E Sodium Lauryl Sulfate	2.0
F Preservative	q.s.

Procedure:

Add the **Veegum** to water slowly, agitating continually until smooth. Mix the sorbitol and glycerin together and use this to wet the sodium CMC. Add B to A. Add C slowly with agitation. Add D and mix. Add E slowly with just enough mixing to obtain a smooth batch. Add F.

Aerosol package:

Concentrate 99%, Propellant 12/114, 60/40 1%

Note:

This formula should be packaged in a Sepro can for maximum evacuation from the aerosol.

No. 6

(Liquid)

A **Veegum**	1.00
Sodium Carboxymethylcellulose (med. visc.)	0.25
Water	21.25
B Sorbitol (70% solids)	12.50
Glycerin	12.50
C Dicalcium Phosphate Dihydrate	50.00
D Flavor	1.00
E Sodium Lauryl Sulfate	1.50
Preservative	q.s.

Procedure:

Dry blend the **Veegum** and CMC. Add to water slowly agitating continually until smooth. Add B and C alternately to A with mixing. Add D and mix until smooth. Add E with slow mixing avoiding incorporation of air.

No. 7

(Anionic)

Lathanol LAL-70 Powder	3.0
Tricalcium Phosphate	26.6
Gum Tragacanth	1.0
Glycerin	45.5
Saccharin	0.2
Water	23.1
Flavor	0.6

Procedure:

SOLUTION I: Disperse the gum in 25 parts of the glycerin. Dissolve the saccharin in the water. Slowly add to gum/glycerin mixture, agitating continually until smooth.

SOLUTION II: Mix **Lathanol LAL-70** with remaining glycerin (use mortar. Hobart mixer, ribbon blender, or similar equipment). Add tricalcium phosphate to **Lathanol LAL-70** and glycerin. Mix until a thick paste is obtained. Add Solution I to Solution II and continue mixing until uniform. Add flavor.

Note:

Tricalcium phosphate may be increased or decreased to obtain desired consistency.

Denture Cleaner

A	**Veegum**	1.0
	Sodium Carboxymethylcellulose (med. visc.)	0.5
	Water	28.9
B	Saccharin	0.1
	Sodium Benzoate	1.0
C	Sorbitol (70% sol'n.)	9.0
	Glycerin	9.0

D	Dicalcium Phosphate (dihydrate)	36.0
	Dicalcium Phosphate (anhydrous)	12.0
E	Flavor	0.5
F	Sodium Lauryl Sulfate	2.0

Procedure:

Dry blend the **Veegum** and the CMC. Add to the water slowly, agitating continually until smooth. Add B to A and mix. Blend C and add to A and B. Blend D and add to the mixture. Add E and F, one at a time, to other components and mix until uniform.

Denture Adhesive

(Nonionic)

A	**Vinac B-15**	46
	Ethyl Alcohol	32
B	**Veegum**	1
	Water	18
C	**Atmos 300**	3

Procedure:

Add the polyvinylacetate (**Vinac B-15**) slowly with stirring to the alcohol which has been heated to 65-70 C. Maintain temperature and stir until all the polyvinylacetate has dissolved and a clear gel is formed. (Maintain the alcohol level by replacing any alcohol that has boiled off.) Add the **Veegum** to the water slowly, agitating continually until smooth. Add B to A with agitation. Stir until smooth. Add C and stir until uniform.

Denture Cleaner Tablet

	grams
A **Veegum WG**	5
Sodium Perborate	13
Tetrasodium Pyrophosphate (anhydrous)	25
Sodium Chloride	13
Tartaric Acid	9

Sodium Phosphate (dibasic)	12
Citric Acid	7
Sodium Bicarbonate	16
B Isopropyl Alcohol	25
C **Veegum WG**	2

Procedure:

Granulate A with B. Pass through a No. 10 mesh screen. Dry the granulation 1 h at 105 C and pass through a No. 16 mesh screen. Dry blend with C and compress.

Directions for use:

Add one tablet to a glass of hot water. Soak dentures in solution one-half hour or overnight and rinse.

Chapter II

HOUSEHOLD CLEANERS

Carpeting and Upholstery Cleaner

Formula No. 1

(Anionic)

Sipex 206-C	39
Butyl Carbitol	3
Formaldehyde	0.1
Water (soft)	58
Perfume	q.s.

No. 2

(Anionic)

Sipex 206C-SP	42.3
Water (soft)	57.6
Formaldehyde	0.1
Perfume	q.s.

No. 3

(Nonionic)

Crodasinic LS35	1.00
Sodium Lauryl Sulfate	1.00
Methofas 6S HPM20	0.40
Volpo 04	0.50
Perchloroethylene	7.50

Water (deionized)	77.10
Propellant 12	12.50

Procedure:

Dissolve **Methofas** in part of the water previously heated to 65 C, stir until fully hydrated, add to this the remaining 'cold' water and stir. Add the surface active agents, followed by the perchloroethylene and stir until homogeneous; then fill into containers and introduce propellant.

Aerosol Rug Shampoo

Formula No. 1

(Nonionic-Anionic)

Igepon AC-78	9.0
Gantrez® AN-119 Copolymer	1.0
Monamide S	1.0
Water	89.0

Adjust to pH 8.0 with sodium hydroxide.

Fill:

90% of above and 10% isobutane or 10% F11/F114 (50/50)

Note:

This formulation can also be used in a liquid rug shampoo at an 8:1 dilution.

No. 2

Crodasinic LS35	26.00
Crillon LDE	2.00
Triethanolamine Lauryl Sulfate	4.00
Water (deionized)	68.00
Propellant 12/114 (60:40)	

Procedure:

Combine detergents in water by warming and stirring. Cool and fill into containers. Introduce propellant.

Rug Shampoo Concentrate

Formula No. 1

(Anionic)

		Mixing Order
Stepanol RS	35.0	3
Stepanol Mg	15.0	4
Stepan LIPA	1.0	2
Water	49.0	1

pH: 8.0 ± 0.5. If necessary, adjust with H_2SO_4 to lower pH; NaOH to raise pH.
Gardner Color: 1
Brookfield Visc.: About 15 cps @ 25 C
Appearance: Light amber liquid.

Procedure:

Combine water and **Stepan LIPA**. Heat (70 C) and stir until **Stepan LIPA** is uniformly dispersed. Add other ingredients and stir until uniform. Adjust pH.

No. 2

(Anionic, 20% Active)

Alkasurf WAQ	28.5
Alkasurf SS-L9ME	28.5
Fomaldehyde	0.2
Water	42.8
Color and Perfume	q.s.

Procedure:

Warm the above mixture to 100 F to effect a clear product. Cool to 80 F, add formaldehyde, perfume and color, if desired.

Dilute to any required concentration.

No. 3

(Anionic)

Monamate CPA (40%)	10.0

| Monawet SNO-35 | 12.0 |
| Water | 78.0 |

Note:
Dilute 1:7 with water for use application. This formulation dries to a crisp, friable residue. If a more brittle residue is desired a few percent styrene maleic anhydride resin may be added.

No. 4

(Anionic)

Conco Sulfate TL-LOW	20.0
Conco AAS 60 S	20.0
Water	60.0

No. 5

(Anionic)

Conco Sulfate WA	30.0
Isopropanol	10.0
Water	60.0

No. 6

(Liquid)

Crodasinic LS35	37.60
Sodium Lauryl Sulfate	20.00
Water (deionized)	42.00
Formalin	0.40

Procedure:
Combine detergents in water and stir until dissolved.

Note:
This formulation could be diluted 50:50 with water to form the product which should then be further diluted by approximately 1 part to 20 parts of water before mechanical application.

Rug Shampoo

Formula No. 1

Dispal 'Alumina' Solids (in acid dispersion)	22.00
Sodium Lauryl Sulfate	10.00
Crodasinic LS35	24.00
Water (deionized)	44.00

Procedure:

Combine the detergents and water and disperse the **Dispal** into the solution. This preparation may be packaged in pressurized or nonpressurized containers. In either case they should be shaken before use.

When packed in pressurized containers (aerosols) the following product/propellant ratio is satisfactory.

90% concentrate
10% isobutane

No. 2

(Anionic)

Monawet SNO-35	12.5
Monamate CPA (40%)	12.5
Water	75.0

Note:

Dilute 1:7 with water for use application. This formulation dries to a crisp, friable residue. If a more brittle residue is desired a few percent styrene/maleic anhydride resin may be added.

No. 3

(Anionic)

Conco Sulfate TL-LOW	20.0
Conco AAS 60 S	20.0
Water	60.0

No. 4

(Anionic)

Conco Sulfate WA	30.0
Isopropanol	10.0
Water	60.0

No. 5

(Nonionic)

Water	77.00
TSPP	0.65
Sodium Lauryl Sulfate (30%)	9.20
TEA Lauryl Sulfate (40%)	9.00
ESI-Terge B-15	3.00
Butyl Cellosolve	1.00
Versene 100	0.15

Procedure:
Dissolve TSPP in water with adequate agitation. When solution clears add other ingredients as shown.

No. 6

(With Solvent)

Crodasinic LS35	12.00
Sodium Lauryl Sulfate	8.00
Crillon LDE	1.00
Isopropyl Alcohol	10.00
Water (deionized)	69.00

Procedure:
Combine detergents and water, add isopropyl alcohol (adjust percentages as required); stir until homogeneous solutions results.

Commercial "Spray-Vac" Rug Cleaner

(Anionic-Amphoteric, Low Foam)

Water	78.35
Tetrapotassium Pyrophosphate	3.35

EDTA, Na$_4$ (40% sol'n.)	0.50
Monateric LF Na-50	12.80
Monawet SNO-35	5.00

Suggested use level: 2 oz per gal; pH (as is) approx. 11.5 – pH (use conc.) approx. 9.6.

Note:

Monawet SNO-35 is incorporated because of its unique wetting properties. At application temperatures (approx. 140 F) it provides excellent initial wetting, but on cooling to room temperature it loses its wetting properties which permits the rug to dry faster.

All-Purpose Cleaner

Formula No. 1

(Anionic)

Nacconol 40F	5
Sodium Sesquicarbonate	50
Sodium Tripolyphosphate (hexahydrate)	25
Trisodium Phosphate (hydrate)	20
Color and Perfume	as desired

No. 2

(Anionic, Liquid)

Water	83.0
N-Silicate	10.0
Secondary Alcohol + 9 E.O.	4.0
Monafax 872	3.0

No. 3

Water	16.0
TKPP (60%)	65.0
Monafax 872	14.0
Nonylphenol + 9.5 E.O.	5.0

No. 4

(Anionic, Liquid)

Water	21.7
TKPP (60%)	58.3
Monafax 872	15.0
Sodium LAS (45%)	5.0

No. 5

(Anionic, Liquid)

Water	48.0
Nonylphenol + 9.5 E.O.	2.0
Metso (anhydrous)	12.0
TKPP (60%)	20.0
KOH	12.0
Monafax 872	6.0

No. 6

(Anionic, Liquid)

Hot Water	50
Questex® 45 W	1
LAS Type Surfactant	15
Victawet 35B	2
Sodium Xylene Sulfonate	6
TKPP (60% sol'n.)	25

No. 7

(Anionic, Liquid, Heavy-Duty)

Hot Water	40
Questex®	1
KOH (solid)	0.5
Carboxy Methyl Cellulose	0.5
Biosoft D-60	10

Victawet 35B	2
Sodium Xylene Sulfonate	7
Cold Water	11.6
TKPP (60% sol'n.)	20.0
Tinopal UNPL	0.4
Sodium Silicate Solution	7

No. 8

(Anionic)

		Mixing Order
Steol CS-460	20.0	6
Stepanate X	15.0	2
Bio Soft S-100	6.7	4
Tetrapotassium Pyrophosphate	20.0	5
Potassium Hydroxide (45%)	2.8	3
Water	35.5	1

pH: 11.9
Gardner Color: 1
Brookfield Visc.: About 30 cps @ 25 C
Appearance: Clear light yellow.

Procedure:

Combine water, **Stepanate X** and potassium hydroxide. Add the **Bio Soft S-100** with stirring. Adjust pH to 7-8. Add **Steol CS-460** and the phosphate and stir until uniform.

Note:

This formulation can also serve as a base for a heavy duty liquid laundering detergent to which carboxymethyl cellulose, optical brightener, etc., should be added.

No. 9

Nacconol 40 F	45.0
Tallow Soap	5.0
Sodium Tripolyphosphate (anhydrous)	41.0

Sodium Metasilicate (anhydrous)	7.5
Carboxymethylcellulose	1.0
Optical Brightener	0.5
Perfume	as desired

No. 10

(Nonionic, Pine-Type)

Sodium Dodecyl Benzyl Sulfonate (low cost)	11.5
Triton X-102	5.0
Clindrol 200-L	2.5
Sodium Tripolyphosphate	1.0
Pine Oil	3.0
Optical Brightener	0.1
Water	76.9

Note:
Perfume and color as desired can be added to the above formulation.

No. 11

(Nonionic)

Volpo T8	8.00
Volpo T15	8.00
Water (deionized)	83.80
Formalin	0.20

Procedure:
Simply dissolve the **Volpos** in water by warming and stirring.

Note:
A detergent reodorant may be produced by the inclusion of pine oil. For solubilization of pine oil in these systems **Volpo T15** is recommended.

No. 12

(Nonionic, Spray-On)

		Mixing Order
Makon 10	3.0	3
Trisodium Phosphate	2.0	2

| Butyl Cellosolve | 5.0 | 4 |
| Water | 90.0 | 1 |

pH: 11.9.
Brookfield Visc.: About 5 cps @ 25 C.
Appearance: Colorless liquid.

Procedure:
Dissolve TSP in water. Add **Makon 10** and butyl cellosolve. Stir until uniform.

No. 13

(Nonionic)

A	**Veegum® HS**	4.00
	Kelzan	0.50
	Water	68.50
B	**Monamid 150-ADD**	0.50
	Plurafac C-17	2.50
	Tetrapotassium Pyrophosphate	1.25
	Potassium Phosphate (tribasic)	0.75
	Water	21.00
C	Ammonium Hydroxide (28%)	1.00

Procedure:
Prepare A by dry blending the **Veegum HS** and **Kelzan** and adding slowly to the water, agitating continually until smooth. Combine B, stirring slowly to dissolve the phosphates (avoid incorporation of air). Add B to A with mixing. Add C and mix until uniform.

Packaging:
This formula is a creamy liquid and can be packaged in a plastic squeeze bottle.

Comments:
In this formula the superior electrolyte stability of **Veegum HS** recommends its use as a thickening agent, along with the **Kelzan**, to form a stable liquid cream; the potassium phosphates function as the soluble detergent builders. **Plurafac C-17** and **Monamid 150-ADD** are both nonionic, biode-

gradable surfactants which work together in this composition to provide detergency and which are stable in use with the active ingredient, ammonium hydroxide. This formula offers desirable grease cutting and soil removal properties and is especially convenient for use on vertical surfaces.

No. 14

(Nonionic-Anionic)

Ninol 1301	10
Tetrapotassium Pyrophosphate	5
Propylene Glycol	5
Bio Soft D-60	5
Pine Oil	0.1
Water	balance

No. 15

(Nonionic-Anionic)

Ultra Blend® 100	4
Ultra® SXS Liquid (40%)	3
Tetrapotassium Pyrophosphate	3
Crystalline Trisodium Phosphate	4
Butyl Cellosolve	3
Water	83

Procedure:

Add the water first, then add the components as listed, mixing until clear after each addition.

No. 16

(Nonionic-Anionic)

		Mixing Order
Ninol 1301	10.0	5
Tetrapotassium Pyrophosphate	5.0	2
Propylene Glycol	5.0	4

Bio Soft D-60	5.0	6
Pine Oil	0.1	7
Stepanate X	10.0	3
Water	64.9	1

Procedure:
Dissolve the phosphate in the water. Add the other ingredients in the order listed with good agitation.

No. 17

(Nonionic-Anionic)

Alkamide 2204	5.0
Alkasurf LA-Acid	1.5
Sodium Tripolyphosphate	3.0
Trisodium Phosphate Fines	4.0
Alkatrope SX-40	5.0
Water	73.5
Monoethanolamine	3.0
Butyl Cellosolve	5.0

Note:
The above formulation is an excellent all-purpose cleaner and doubles up as an extremely effective nonammoniacal floor cleaner and wax stripper. This formulation can be modified to suit a particular requirement.

No. 18

(Nonionic-Anionic)

		Mixing Order
Sodium Metasilicate·5H$_2$O	21.25	2
Na$_4$ EDTA	8.50	3
Ninol 1285	5.00	4
Steol KS-460	7.50	5
Stepanate X	2.00	6
Water	55.75	1

Procedure:
Blend deionized water, sodium metasilicate and Na$_4$EDTA. Mix thoroughly. Then add the **Ninol 1285, Steol KS-460** and **Stepanate X.**

No. 19

(Nonionic-Anionic)

		Mixing Order
Stepanate X	25.0	3
Ninol 1281	3.0	5
Tetrapotassium Pyrophosphate	10.0	2
NH$_4$OH (28%)	2.0	4
Latex E-295	1.0	6
Water	59.0	1

Procedure:
Dissolve the phosphate in most of the water. Add the **Stepanate X**, NH$_4$OH, and **Ninol 1281** and mix. Add the **Latex E-295** diluted with the remainder of the water. Mix.

No. 20

(Nonionic-Anionic)

		Mixing Order
Bio Soft D-40	10.0	2
Ninol 1285	7.0	4
Bio Soft S-100	3.6	6
Isopropylamine	1.4	5
Stoddard Solvent	5.0	7
Isopropyl Alcohol	10.0	3
Water	62.0	1
Pine Oil	1.0	8

Procedure:
Combine ingredients in sequence, mixing thoroughly after each addition until clear and uniform.

No. 21

(Nonionic-Anionic)

		Mixing Order
Bio Soft D-40	10	2
Ninate 411	5	5
Ninol 1285	7	4
Stoddard Solvent	5	6
Isopropyl Alcohol	10	3
Water	62	1
Pine Oil	1	7

Procedure:
Combine ingredients in sequence and stir until uniform.

No. 22

(Nonionic-Anionic)

Water	89.00
Trisodium Phosphate	3.00
Sodium Tripoly Phosphate	3.00
ESI-Terge HA-20	5.00

Procedure:
Add in order listed with adequate agitation. Allow all powders to dissolve before adding **ESI-Terge HA-20**. Agitate until clear.

No. 23

(Nonionic-Anionic)

		Mixing Order
Bio Soft D-40	15	2
Ninol 1285	7	4
Stoddard Solvent	5	5

Isopropyl Alcohol	10	3
Water	62	1
Pine Oil	1	6

Procedure:
   Combine ingredients 1 through 4 in sequence, mixing thoroughly until clear after each step. Using moderate agitation, add 5 and 6 and mix until solution is uniformly hazy. The product can be clarified at either a pH of about 7.3 or pH of about 9.0.

Note:
   For pH of 7.3: Slow add 50% H_2SO_4 until target pH is obtained.
   For pH of 9.0: Add sodium sulfate in small increments until permanent clarity is obtained (usually requires about 0.5 g sodium sulfate).

No. 24

(Nonionic-Amphoteric)

Water	81.84
Metso (anhydrous)	4.50
TKPP	9.00
Monateric CEM (38%)	1.33
Tergitol 15-S-9	2.00
NaLAS (100%)	1.33

Note:
   Use concentration: 1–4 oz/gal.
   This formula contains TKPP as a partial replacement for the nonionic ethoxylate, and has excellent stability, detergency and foaming properties.

No. 25

(Anionic-Nonionic-Amphoteric)

Water	86.64
Metso (anhydrous)	6.00
Monateric CEM (38%)	1.36
Tergitol 15-S-9	4.00
Monamine ALX-100S	2.00

Procedure:

Add the components in the order listed. Care should be taken to make sure the silicates are completely dissolved before adding the organic detergents and hydrotropes.

Note:

Use concentration: 1-4 oz/gal.

This formulation has excellent wetting, detergent, rinsing and foaming properties. It is recommended as a medium duty cleaner for truck bodies, floor scrubs, wax strippers, etc. This product is clear from −5 C to 70 C. This formula would require approximately 4 times as much sodium xylene sulfonate as the amount of **Monateric CEM** (38%) used to achieve clarity. Other nonionic ethoxylates, such as those based on primary alcohol or alkyl phenol may be substituted for the **Tergitol 15-S-9**.

No. 26

(Amphoteric, Heavy-Duty)

Monateric Cy Na (50%)	4.0
Sodium Metasilicate · 5H$_2$O	4.0
Sodium Carbonate (anhydrous)	2.0
Nonyl Phenyl Polyethylene Glycol Ether	4.0
Butyl Carbitol	1.0
Water	85.0

No. 27

(Phosphate-Free)

Arylan LQ	20.00
Perlankrol SXS1	10.50
Sodium Heptonate (industrial)	18.00
Triethanolamine	1.00
Water (deionized)	50.50

Note:

The sodium heptonate imparts alkalinity and sequesters calcium and magnesium ions, allowing the wetting agent to become fully active.

No. 28

(Phosphate-Free)

Butyl Cellosolve	4.00
Sodium Silicate	2.00
Nansa HS 85/S	1.00
Crillon LDE	1.00
Crodaquest EDTA	1.00
Water (deionized)	91.00

No. 29

(Liquid)

A **Veegum**	2.0
Water	89.0
B **Super Floss**	1.6
Snow Floss	2.4
C Sodium Hypochlorite (5% sol'n.)	5.0

Procedure:
Add the **Veegum** to the water slowly, agitating continually until smooth. Add B to A mixing thoroughly to disperse abrasive particles. Add C slowly agitating until uniform.

Directions for use:
Spray surface from a distance of about 8 in. Wipe with dry cloth or paper towel. If dispensed from a bottle, pour a small amount of the cleaner on a dry cloth or paper towel and wipe. Rinsing with water is not necessary.

Note:
This cleaner is suitable for use on hard surfaces such as bathtubs, sinks, ceramic tiles, stoves, and refrigerators. It may be packaged in a pump spray dispenser or in a bottle.

Floor Cleaner

Formula No. 1

(Anionic)

Nacconol 90F	5
Trisodium Phosphate (monohydrate)	10
Water	85

No. 2

(Anionic)

		Mixing Order
Stepan HDA-7	5	5
Stepanate X	3	2
Trisodium Phosphate	4	4
Sodium Tripolyphosphate	4	3
Water	84	1

Procedure:
Dissolve the phosphates in the water and **Stepanate X**. Add the **HDA-7** and stir until uniform.

No. 3

(Nonionic-Anionic)

Ninol 1301	7
Tetrapotassium Pyrophosphate	4
Bio Soft D-60	4
Water	85

Procedure:
In production the phosphate would first be dissolved in the water (warmed to about 120 F) then the **Bio Soft D-60** added, and finally the **Ninol 1301**. Since the **Ninol** is a soft wax, it could readily be melted by storing the drum over steam pipes, or by means of an electric strap or im-

mersion heater. If desired, it could easily be scooped out of the open head drums.

The **Bio Soft D-60** is a 50% sodium alkyl aryl sulfonate slurry (in alcohol) used as a coupler to ensure clarity of the formulation. Actually a coupler is not essential, but it helps to prevent separation of the floor cleaner under very cold or hot storage conditions. In this formulation, couplers like xylene sulfonate (**Stepanate X**) are not recommended, as they lower viscosity.

Note:

This floor cleaner exhibits high viscosity, controlled foam and excellent detergency. It can be shipped in unlined steel drums without rusting, although addition of about 0.2% sodium nitrite will further increase rust resistance.

This cleaner can be diluted with hard water, without developing the cloudiness, which many amide based floor cleaners exhibit in hard water areas.

No. 4

(Nonionic–Anionic)

Alkamide 2204	7
Sodium Tripolyphosphate	3
Trisodium Phosphate	2
Alkatrope SX-40	q.s.
Water	to 100

Procedure:

When manufacturing floor cleaners of this type, the simplest procedure is to dissolve the phosphates in warm water followed by the **Alkamide 2204**. The antirusting properties of this formulation permit plain steel tanks to be used, and plain open-head steel drums are often used as mix tanks for small quantities.

The **Alkatrope SX-40** shown above is a coupler and may be necessary in some formulations of higher salt levels. Since **Alkamide 2204** itself has excellent solubility and compatibility with phosphate and silicates, a coupler is not essential. However many compounders may consent to its use to meet certain cold temperature and freeze-thaw characteristics.

This formulation will perform well even in hard water areas.

Directions for use:

For general use, it is recommended that the above formulation be diluted 4 oz/gal. For stripping off floor wax it is recommended the cleaner be only diluted about five times and applied hot, then allowed to soak in for about 15 min.

Note:

The above formulation has a pH of about 10. If a lower pH of about 9.5 is desired, then only sodium tripolyphosphate should be used. Cleaners of this type are widely used for floors and walls in office buildings, hospitals and schools. The moderate alkalinity (pH 10 or less) makes it safe for all types of tile and painted surfaces.

For most effective wax stripping, 5-10% ammonium hydroxide can be added to the above formulation. If the ammonia odor is considered objectionable the ammonia could be substituted with 5% monoethanolamine or morpholine.

No. 5

(Nonionic-Anionic)

		Mixing Order
Ninol 1281	7.0	5
Sodium Tripolyphosphate	2.0	4
Trisodium Phosphate	2.0	3
Stepanate X	3.0	2
Water	86.0	1

Procedure:

Dissolve STPP and TSP in water and **Stepanate X**. Add **Ninol 1281**. Mix thoroughly.

No. 6

(Nonionic-Anionic)

		Mixing Order
Ninol 1301	7	4
Bio Soft D-60	4	3

Tetrapotassium Pyrophosphate	4	2
Water	85	1

Procedure:

Dissolve the phosphate in the water. Add **Bio Soft D-60** and **Ninol 1301** (warmed to melt).

Wash and Wax Floor Finish

Neo Cryl A247H	65.10
AC-629	14.00
Shanco 334 or **Durez 19788** or **SR-88**	14.00
Carbitol	1.85
Tri Butoxyethyl Phosphate	0.45
Detergent Composition*	4.60

*Detergent Composition

Miranol® L2M-SF Conc.	11.0
Sodium Metasilicate Pentahydrate	3.5
Water	85.5

Liquid Tile Cleaner

Formula No. 1

(Nonionic-Anionic)

A **Veegum T**	1.5
Water	71.5
B **Darvan No. 7**	2.0
Triton X-102	5.0
Sulframin 85	5.0
Pine Oil	5.0
C Mild Abrasive	10.0

Procedure:

Add the **Veegum T** to the water slowly, agitating continually until smooth. Add the components in B to A in the order listed with mixing. Add C to A and B slowly with stirring. Mix until uniform.

Aerosol Package:

Concentrate 91%; Propellant 12 9%

No. 2
(Nonionic-Anionic)

		Mixing Order
Bio Soft D-40	10.0	2
Ninol 2012 Extra	3.0	3
Sequestrene NA-4	1.0	4
Methyl Paraben	0.1	5
Water	85.9	1

Procedure:
Combine ingredients in sequence mixing thoroughly after each addition until clear and uniform.

Toilet Bowl Cleaner

Formula No. 1

(Anionic)

		Mixing Order
Nacconol 40DBX	4.0	3
Sodium Sesquicarbonate (L.D.)	16.5	2
Sodium Bisulfate	73.5	1
Sodium Dichloroiosocyanurate	2.0	6
Santophen #1	1.5	4
Polyethylene Glycol 400	1.5	5
FD&C Blue No. 1	0.075	8
Fragrance	0.675	7
Cab-O-Sil M-5	0.25	9

Procedure:
Combine (1), (2), and (3); mix 5 min. Dissolve (4) in (5) and administer as atomized spray to dry mix. Mix 10 min after completion of addition. Add (6), mix 10 min; then add (7), (8), and (9) mixing about 5 min after addition of each ingredient.

Equipment:
Auger, plow-share, or twin-cone mixer, ribbon blender, roller mill or equivalent.

No. 2

(Nonionic, Acid)

		Mixing Order
Hydrochloric Acid (37% HCl)	20	2
Water	60	1
Makon	20	3

Procedure:
Combine water and hydrochloric acid. Add the **Makon** and stir until uniform.

Brookfield Visc.: About 3000 cps with M-6; about 300 cps with M-8 @ 25 C.

Note:
Amount of **Makon** can be increased or decreased with a corresponding increase or decrease in viscosity. **Makon 6** gives most viscosity and a white opaque product. Higher Makons give clear product with less viscosity.

No. 3

(Nonionic, Acid)

A **Veegum HS**	1.0
Kelzan	0.5
Water	86.5
B Phosphoric Acid	10.0
C **Triton X-100**	2.0

Procedure:
Dry blend the **Veegum HS** and **Kelzan** and add slowly to the water, agitating continually until smooth. Add B and then C with mixing after each addition and mix until smooth.

No. 4

(Nonionic)

Section 1:

Orthodichlorobenzene	10
Emulsifier 210-70	10
Hydrochloric Acid (24%, 16 Be.)	965

Procedure:
Dissolve the orthodichlorobenzene in the Emulsifier 210-70. Add this solution to the well-stirred 24% hydrochloric acid.

Section 2:

E-153 Latex	5
Water	10
Makon 10 or **Igepal CO-630**	0.4

Procedure:
Dissolve the **Makon 10** (or **Igepal CO-630**) in water and add the **E-153**. Then add the diluted **E-153** slowly, with good agitation, to the hydrochloric acid/orthodichlorobenzene emulsion.

The completed formulation is stable both at room temperature and on oven aging (125 F).

Warning:
This formula, because it contains orthodichlorobenzene, will soften polyethylene containers and should be packaged only in glass bottles.

*Substitute for **Igepal CO-630** are:

Tergitol NPX
Renex 698
Polytergent B-300
Surfonic N-95
Triton X-100
Makon 10

No. 5

(Nonionic-Anionic, Liquid)

		Mixing Order
Bio Soft S-100	2.0	5
Makon 8	3.0	3
Hydrochloric Acid (22 Be)	20.0	4
Sequestrene AA	1.0	6
Sodium Bisulfate	4.0	2
Water	70.0	1
Colorant and Fragrance	q.s.	—

Procedure:

Dissolve 2 in 1 and add 3; mix well and add 4. Add 5 slowly to completely dissolve. Mix 5 min and add 6; stir until dissolved. Perfume and color as desired.

No. 6

(Nonionic-Cationic)

Hyamine 3500 (50% sol'n.)	1.5
Triton N-101	12.5
Triton X-114	12.5
E-153 Latex (50% solids)	19
Water	228
Hydrochloric Acid (32%–20 Be)	1150

Procedure:

With stirring, add the nonionic emulsifiers to the water. Then with continued stirring add the **E-153 Latex**. Follow with the addition of the hydrochloric acid and finally add the **Hyamine** with stirring.

Note:

The formulation given above has been tested by the AOAC Use Dilution Procedure against *Salmella Choleraesius* and *Staphylococcus Aureus*. At a dilution of 2 oz per 3 quarts of water (average toilet bowl volume), the formulation will support disinfectant claims.

Comments:

E-153 should not be used to opacify bowl cleaners which contain cationic surfactants. The stabilizing nonionic emulsifier must be water soluble and acid stable. It is necessary that the E-153 be stabilized with the nonionic emulsifier before addition of the acid. Preferably, the nonionic is dissolved in the water used to dilute the E-153 before addition of the acid. If a more opaque product is prepared by increasing the concentration of E-153, it is necessary to increase the amount of nonionic proportionally. Some nonionics may be less efficient than others in which case a higher concentration may be required in order to obtain a stable product. Low temperature stability of E-153 opacified hydrochloric acid formulas is perfectly satisfactory.

No. 7

(Amphoteric, Acid)

Water	82.5-72.5
Muriatic Acid	15.0-25.0
Monateric CEM (38%)	2.5

Use concentration: 4-8 oz per toilet bowl.

No. 8

(Amphoteric, Acid)

Miranol® C2M-SF Conc.	3.0
Hydrochloric Acid (30%)	20.0
Rodine 213	0.5
Perfume and Color	q.s.
Water	76.5

No. 9

A	**Veegum HS**	1
	Water	92
B	Citric Acid	4
C	Phosphoric Acid	3

Procedure:
Add the **Veegum HS** slowly to the water, agitating continually until smooth. Add B and then C with mixing after each addition and mix until smooth.

Liquid Household Drain Cleaner

(Nonionic-Anionic)

Water	89.5
Sodium Hydroxide (flakes)	9.5
Monaterge 85	1.0

Procedure:
Add ingredients in the order listed with good agitation.

Porcelain Cleaner

(Nonionic)

Water	76.00
TSP	2.00
ESI-Terge B-15	5.25
Tall Oil Fatty Acid	1.75
Kaopolite SF	15.00

Procedure:
Dissolve TSP in water. When solution clears up add **ESI-Terge B-15** and mix well until a good emulsion is formed. Add tall oil fatty acid and mix until emulsion clears. Add **Kaopolite SF** and mix until homogeneous blend is formed.

Cleaner–Sanitizer

Formula No. 1

(Nonionic-Cationic)

Alkasurf LA-16	2-5
Alkamidox CMC-5	1-2
Alkaquest EDTA	0.1-0.2

Sodium Tripolyphosphate	1-3
Alkaquat DMB-451	3-5
Color and Perfume	q.s.
Water	to 100

No. 2

(Cationic, Aerosol)

Solvay Cleanser 600	3.00
Hyamine #1622	0.05
Isopropyl Alcohol	10.00
Water	86.95

Procedure:
Mix all ingredients together until uniform and package in aerosol container.

Pine Oil Disinfectant

	Formula No. 1	No. 2
Yarmor Pine Oil	80.00	65.00
Tall Oil Fatty Acids	9.25	23.5
Caustic Soda	1.25	3.15
Water	9.50	8.35

Procedure:
The above disinfectants may be prepared at room temperature. The fatty acids and pine oil are first blended. A solution of caustic soda in all the water is gradually added to this blend with efficient stirring, which is continued for about 30 min. The respective products should be clear, amber solutions.

Pine-Oil Deodorant

Pine Oil	90 fl. oz.
Rosin FF	40 oz.
Caustic Soda (sol'n.–sp. gr. 1.275 at 25 C)	20 fl. oz.

Procedure:
Dissolve the rosin in the pine oil, heated in a steam-jacketed kettle at 80 C. When all is dissolved, add the caustic soda solution, stirring briskly and maintaining the temperature at 80 C.

Disinfectant Cleaner

Formula No. 1

(Nonionic)

		Mixing Order
Ninex 303		
Ninex 303	7.5	4
Ninol 1281	6.5	5
Dowacide A	5.0	6
Tetrasodium Pyrophosphate	2.0	2
Trisodium Phosphate	2.0	3
Water	77.0	1

Procedure:
Dissolve the phosphates in the water and **Ninex 303**. Add the **Ninol 1281** and **Dowacide A**. Stir. Heating will facilitate the mixing.

No. 2

(Nonionic)

Ninol 1301	10
Quaternary Germicide	15
Tetrapotassium Pyrophosphate	4
Water	71

Note:
This gives a clear, viscous formulation with good cleaning power. Using **Hyamine 1622** as the quaternary, the phenol coefficient was found to be 25.

No. 3

(Nonionic, Pine-Type)

Condensate PS	15.0
Pine Oil	3.0
Water	82.0

No. 4

(Nonionic-Cationic)

		Mixing Order
Ninol 1301	10	4
Hyamine 1622	15	3
Tetrapotassium Phosphate (TKPP)	4	2
Water	71	1

Procedure:

Dissolve **Hyamine 1622** and TKPP in water; add **Ninol 1301.** Stir until uniform.

Copper Cleaner

Formula No. 1

(Nonionic)

A	Veegum K	2.07
	Kelzan	0.23
	Water	78.55
B	Snow Floss	13.60
C	Buffer Solution*	q.s.
D	Triton X-102	4.65
	Lauryl Thioglycolate	0.90
	Perfume and Preservative	q.s.

---

*Buffer sol'n.: 1.46 parts–1M H_3PO_4
                1 part–125 g/l Na_3PO_4

Procedure:

Dry blend **Veegum** and **Kelzan** and add to the water slowly, agitating continually until smooth. Add B to A gradually. Mix until smooth, then buffer this mixture with buffer solution C to a pH of 2.5. Mix components in D until a clear solution is formed. Special care should be taken to avoid incorporation of air. Add D to other components very slowly and mix until uniform.

Directions for use:

Apply copper cleaner with damp cloth. Rinse and dry. Polish with a clean dry cloth.

No. 2

(Nonionic)

A	**Veegum K**	2.07
	Kelzan	0.23
	Water	78.55
B	**Snow Floss**	13.60
C	Buffer Solution*	q.s.
D	**Triton X-102**	4.65
	Benzotriazole	0.90
	Perfume and Preservative	q.s.
	Color	q.s.

*Buffer sol'n.: 1.46 parts−1 M H_3PO_4
1 part−125 g/l Na_3PO_4

Procedure:

Dry blend **Veegum K** and **Kelzan** and add to the water slowly, agitating continually until smooth. Add B to A gradually. Mix until smooth, then buffer this mixture to a pH of 2.5. Mix components in D until a clear solution is formed. Special care should be taken to avoid incorporation of air. Add D to other components very slowly and mix until uniform.

No. 3

Sodium Hydroxide	73.55
Sodium Metasilicate (anhydrous)	15.0

Sodium Gluconate	5.0
Sodium Tripolyphosphate	5.0
Alkylamine Polyglycol (condensate)	0.75
2-Mercaptobenzothiazole	0.7

Procedure:

Mix all ingredients together until uniform.

Dissolve in water to caustic concentration of between 0.5 to 5% (16 to 160 times dilution).

Stainless Steel Cleaner

Sodium Hydroxide	79
Tetrasodium Pyrophosphate	12
Sodium Gluconate	5
Alkylamine Polyglycol (condensate)	3
2-Mercaptobenzothiazole	1

Procedure:

Mix all ingredients together until uniform.

Dissolve in water to caustic concentration of between 0.5 to 5% (16 to 160 times dilution).

Metal Cleaner

(Nonionic)

Kerosene	54
Makon 4	9
Makon 8	9
Phosphoric Acid (75%)	8
Water	20

Procedure:

Mix the **Makons** and kerosene. Combine the acid and water and add to the kerosene with good agitation.

Aluminum Cleaner

Formula No. 1

(Anionic, Liquid Acidic)

		Mixing Order
Bio Soft S-100	3.00	5
Phosphoric Acid (85%)	12.00	4
Butyl Cellosolve	4.00	6
Citric Acid	3.00	2
Sodium Chromate	0.05	3
Water	77.95	1
Fragrance and Colorant	q.s.	7

Procedure:
Combine 1, 2, and 3; mix until clear. Add 4, 5, and 6 in sequence mixing 2 to 3 min after each addition. Perfume and color as desired.

No. 2

(Anionic)

Nacconol 40F	5
Sodium Metasilicate (pentahydrate)	80
Tetrasodium Pyrophosphate	15

No. 3

(Amphoteric)

Miranol® C2M-SF Conc.	5.0- 2.5
Butyl Cellosolve	6.0- 6.0
Phosphoric Acid (85%)	38.0-38.0
Hydrofluoric Acid (52%)	8.0- 8.0
Ethylenediamine Tetra-Acetic Acid	1.0- 1.0
Water	42.0-44.5

No. 4

(Amphoteric, Liquid)

Water	47.0
Phosphoric Acid (75%)	40.0
Hydrofluoric Acid (52%)	6.0
Monateric CEM (38%)	2.0
EDTA, Na$_4$ (40%)	1.0
Butyl Cellosolve	4.0

Note:
Use concentration: 1–2 oz/gal

No. 5

(Amphoteric, Liquid)

Water	47.0
Phosphoric Acid (75%)	40.0
Hydrofluoric Acid (52%)	6.0
Monateric CEM (38%)	2.0
EDTA, Na$_4$ (40%)	1.0
Butyl Cellosolve	4.0

Note:
Use concentration: 1–2 oz/gal

Chrome and Aluminum Cleaner/Polish

(Nonionic)

A	**GE Viscasil® 10,000**	4.0
	Carnauba Wax No. 2	3.0
	Mineral Spirits	57.0
B	**Thixcin R**	4.0
	Mineral Spirits	12.0
C	**Super Floss**	20.0

Procedure:
1. Disperse the **Thixcin R** in Part B with mineral spirits to form a 25% dispersion (paste).
2. Heat Part A to 70 C (158 F) until all the wax is melted. Higher temperatures should be avoided due to possible detrimental effects to the **Thixcin R**. Continue high-shear agitation while adding Part B.
3. Add Part C slowly with high-speed agitation. The material will gel rapidly during this addition period to form a soft paste. Continue agitation until the temperature is 45 C (113 F), otherwise the **Thixcin R** may seed out.

Application:
Apply the polish with a damp cloth, rubbing the surface vigorously. Allow to dry to a white haze and buff with a clean, dry cloth.

Warning:
When solvents are used, as described above, proper safety precautions must be observed. All solvents must be considered toxic and should be used only in well ventilated areas. Prolonged exposure to solvent vapors must be avoided. If flammable solvents are used, storage, mixing, and use must be in areas away from open flames or other sources of ignition. The selection of any solvent, particularly chlorinated hydrocarbon solvents, will require consideration of applicable OSHA, EPA and other federal, state and local regulations.

Liquid Dishwashing Detergent

Formula No. 1

(Anionic)

		Mixing Order
Bio Soft S-100	22.5	4
Steol CA-460	20.0	7
Stepanate X	14.7	3
Stepan LMMEA	5.9	5
NaOH (50%)	5.5	2
Urea	3.0	6
Water	28.4	1

Procedure:
Combine the water, NaOH and **Stepanate X**. Add the **Bio Soft S-100** with constant stirring. Melt the **Stepan LMMEA** and add to the warm neutralization product. Adjust pH and add the **Steol CA-460** and urea.

No. 2

(Anionic)

		Mixing Order
Bio Soft S-100	24.52	4
NaOH (50%)	6.35	2
Steol CA-460	12.90	6
Stepanate X	18.24	3
Stepan P-621	2.53	7
Water	32.46	1
Alcohol SD 3A	0.50	5
Morton Latex E-295	2.05	8
Citric Acid	0.45	9

Procedure:
Combine most of the water, the NaOH and **Stepanate X**. Neutralize with the **Bio Soft S-100**. Add the **Steol CA-460** and mix until uniform. Add the 3A alcohol and **Stepan P-621** and stir. Adjust the pH to 6.5–7.0. Add the **Latex E-295** diluted with the balance of the water.

No. 3

(Anionic)

Alkasurf LA Acid	15
Alkasurf ES 60	6
Caustic Soda Flake	2
Alkamide CDO	4
Alkasurf SX 40	20
Water	to 100
Perfume, Color, Opacifier	q.s.
Phosphoric Acid	to pH 7–7.5
Total Solids	32%

Procedure:

To water and caustic soda, add the **Alkasurf LA Acid** slowly with agitation. Add the remaining ingredients and adjust to pH 7.0-7.5. Should opacifier be required, follow manufacturers' directions regarding method of addition.

No. 4

(Nonionic)

Water	82
Detergent Concentrate 840	15
Monamid 716	3
Phosphoric Acid (75%)	to pH 7

Note:

This formulation has good viscosity and excellent flash foam, and foam stability in the presence of food soils.

No. 5

(Nonionic)

Sodium Dodecyl Benzene Sulfonate	23
Triton X-102	10
Clindrol 200-L	5
Water	62

Note:

The amounts indicated in the above formula are on a 100% active basis.

No. 6

(Nonionic)

Sodium Dodecyl Benzene Sulfonate	20
Water	65
Triton X-102	10
Clindrol Superamide 100LM	5

Note:

For a lotion type dishwash detergent add 2 parts of an opacifying latex to the above formula.

No. 7

(Nonionic-Anionic)

Water	70.00
ESI-Terge T-60	25.00
ESI-Terge S-10	5.00

Procedure:
Add in order listed and agitate until uniform.

No. 8

(Nonionic-Anionic)

Water	72.80
Sodium Hydroxide (50%)	4.40
ESI-Terge DDBSA	16.28
ESI-Terge 320	1.63
Triethanolamine	traces
ESI-Terge S-10	4.89

Warning:
Chemical neutralization continues for hours. Do not pack until the next day. If packing is necessary, do not cap containers tightly.
Protective clothing should be used when handling sodium hydroxide.

	No. 9	No. 10
	(Nonionic-Anionic)	
Alkasurf LA Acid	12	15
Alkasurf ES 60	12	6
Alkasurf SX 40	5	20
Alkamide CDE	3	3
Caustic Soda Flake	1.6	2
Water	to 100	to 100
Phosphoric Acid	to pH 7.0	to pH 7.0
Color, Perfume, and Opacifier	q.s.	q.s.
Preservative	q.s.	q.s.
Total Solids %	25	31

Procedure:

To water and caustic soda, add the **Alkasurf LA Acid** slowly with agitation. Add the remaining ingredients, adjust the pH to 7.0. Should opacifier be required, follow manufacturers' directions regarding the method of addition.

No. 11

(Nonionic-Anionic)

		Mixing Order
Bio Soft D-62	30	1
Bio Terge AS-35CL	34	2
Ninol 128 Extra	3	6
Makon 12	5	5
3A Alcohol	6	3
Water	20	4
Latex E-295	2	7

Procedure:

Combine about half of the water and the other ingredients except the **Latex E-295**. Agitate until uniform. Add the **Latex E-295** diluted with the remaining water and stir until uniform.

No. 12

(Nonionic-Anionic)

		Mixing Order
Bio Terge AS-35CL	73.71	5
Steol CA-460	10.83	6
Ninol 128 Extra	5.00	7
Urea	2.00	4
Magnesium Sulfate	0.80	3
Citric Acid	0.12	2
Water	7.54	1

Procedure:
  Blend together the deionized water, citric acid, magnesium sulfate, urea, **Bio Terge AS-35CL** and **Steol CA-460**. Then slowly with good mixing add the **Ninol 128 Extra**. Care should be taken in adding the **Ninol 128 Extra** that the pH of the blend doesn't go high enough to liberate ammonia from the **Steol CA-460**.
  It may be necessary to add a small amount of **Stepanate X** to obtain clarity of the finished product.

No. 13

(Nonionic-Anionic)

		Mixing Order
Bio Soft S-100	21.80	4
NaOH (50%)	5.40	2
Steol CA-460	10.03	6
Ninol AA-62 Extra	2.62	5
Stepanate X	19.40	3
Latex E-295	0.87	7
Water	37.88	1
NaCl	2.00	8

Procedure:
  Combine the NaOH and **Stepanate X** with most of the water. Neutralize with **Bio Soft S-100**. Add the **Ninol AA-62 Extra** (melted) and stir until homogeneous. Adjust pH. Add the **Steol CA-460**. Add the **Latex E-295** diluted with the remainder of the water.

No. 14

(Nonionic-Anionic)

		Mixing Order
Steol CS-460	44.76	5
Makon 10	6.10	6
Bio Soft S-100	18.40	4

NaOH (50%)	4.93	2
Stepanate X	4.76	3
Water	21.05	1

Procedure:
Combine water, NaOH and **Stepanate X** and neutralize with **Bio Soft S-100**. Add the **Steol CS-460** and **Makon 10**. Adjust pH.

No. 15

(Nonionic-Anionic)

		Mixing Order
Bio Soft D-40	40.0	2
Ninol 2012 Extra	6.0	5
Steol CA-460	10.0	4
Stepanate X	10.0	3
Methyl Paraben	0.1	6
Water	33.9	1

Procedure:
Combine ingredients 1, 2, and 3; mix until homogeneous. Add 4 slowly and mix until completely dissolved. Then add 5 in small portions keeping pH at 7.0–7.5 by adding 10% sulfuric acid. Finally, add 6 and mix until dissolved.

No. 16

(Nonionic-Anionic)

		Mixing Order
Bio Soft S-100	19.4	3
Stepanate AM	18.0	4
Steol CA-460	12.0	5
Ninol AA-62 Extra	4.0	6
Sodium Hydroxide (50%)	5.2	2
Water	41.4	1

Procedure:

Combine water and sodium hydroxide. Add **Bio Soft S-100** with stirring. Add remaining ingredients in order indicated (melt the **Ninol AA-62 Extra** before adding) and stir until a uniform mixture results. Adjust pH.

No. 17

(Nonionic-Anionic)

		Mixing Order
Bio Soft N-300	25	3
Ninol AA-62 Extra	5	2
Water	70	1

Procedure:

Add **Ninol AA-62 Extra** to the water and heat to about 60 C. Stir until **Ninol AA-62 Extra** is melted and dispersed. Add **Bio Soft N-300** and stir until uniform. Adjust pH.

No. 18

(Nonionic-Anionic)

		Mixing Order
Bio Soft S-100	34.1	5
Makon 12	8.7	6
Stepanate X	7.2	4
Urea	11.0	7
NaOH (50%)	9.1	3
Na_2SO_4	2.0	2
Water	27.9	1

Procedure:

Dissolve the Na_2SO_4 in the water. Add the NaOH and **Stepanate X**. Add the **Bio Soft S-100** with constant agitation. Adjust pH. Add the **Makon 12** and urea.

No. 19

(Amphoteric)

Miranol® C2M-SF Conc.	20.0-40.0
Sodium Lauryl Sulfate (28%)	15.0-15.0
Lauric Diethanolamide (high active)	5.0- 5.0
Water	60.0-40.0

Note:

This is a fine hand dishwashing product providing extremely stable foam, completely nonirritating to skin and eyes and completely biodegradable.

In this formulation **Miranol® C2M-SF Conc.** may be replaced by **Miranol® OM-SF Conc.** or **Miranol® L2M-SF Conc.**

Dishwashing Powder

Formula No. 1

(Anionic, Light-Duty)

		Mixing Order
Bio Soft EA-10	6.0	5
Bio Soft S-100	11.1	4
Steol CS-460	4.0	3
Sodium Carbonate (light, granular)	20.9	1
Stepan LMMEA	2.0	5'
Snowlite II	54.0	2
Sodium Silico Aluminate	1.0	6
Cab-O-Sil EH-5	1.0	7

Procedure:

Combine 1 and 2 in mixer; blend 5 min. Atomize 3 into dry blend and then mix for 5 min. Heat 4 to 50 C and atomize into blender onto "dry" substrate. Mix 30-60 min.* (Vent blender to release CO_2 generated.) Combine 5 and 5', heat to 70 C to dissolve 5', and atomize through preheated pipeline into blender. Mix 5 min. Add 6 and 7, mixing 5 min after each addition. Perfume and color as desired; pack out via 10 mesh sieve.

Tamped density = 0.66 g/ml
Product consists of white, nearly-spherical granules.
pH: 1% sol'n. = 9.85
      0.1% sol'n. = 9.90

Equipment:
Rotating drum, twin-shell blender, plowshare mixer, ribbon blender, or equivalent.

---

* Preheating 1 and 2 to 45–50 C initially will reduce total mixing time.

No. 2

(Anionic, Light-Duty)

		Mixing Order
Bio Soft S-100	9.8	6
Sodium Carbonate (light, granular)	19.6	1
Sodium Hexametaphosphate	8.0	7
Sodium Silico Aluminate	2.0	8
Sodium Sesquicarbonate	22.0	2
Sodium Tripolyphosphate (light, granular)	14.0	3
Sodium Sulfate	17.6	4
Sodium Chloride	6.0	5
Water	1.0	6'

Procedure:
Combine 1, 2, 3, 4, and 5; blend 5–10 min. Combine 6 and 6' and heat to ca. 50 C. Atomize 6/6' into dry blend. Heat mixer to ca. 40–50 C to accelerate reaction. Vent mixer. Charge 7, mix 5 min, and add 8. Mix 5 min and pack out via 10 mesh sieve.

Appearance: White, medium coarse granules.
pH: 1% sol'n. = 10.2
      0.1% sol'n. = 10.0
Density (tamped): 0.91 g/ml

Equipment:
Ribbon blender, twin-shell blender, plowshare mixer, or equivalent.

No. 3

(Anionic, Light-Duty)

		Mixing Order
Bio Soft S-100	9.80	9
Steol CS-460	1.75	6
Snowlite II	45.10	1
Sodium Carbonate (light, granular)	10.20	2
Sodium Bicarbonate	3.00	3
Sodium Tripolyphosphate (light, granular)	4.00	4
Monosodium Phosphate (monohydrate)	3.00	7
Sodium Sesquicarbonate	13.85	5
Sodium Sulfate	7.80	8
Water	0.50	4'
Sodium Silico Aluminate	1.00	10

Procedure:
Combine 1, 2, 3, 4, and 5; mix 5 min and atomize 4' into blend. Mix 2-5 min. Atomize 6 into batch; then, add 7 and 8. Mix 5 min. Atomize 9 into batch, mix 5-10 min and add 10. Mix until uniform and pack out via 10 mesh sieve.

Appearance: White, medium coarse granules.
pH: 1% sol'n. = 9.9; 0.1% sol'n. = 9.8
Density (tamped) = 0.85 g/ml

Equipment:
Ribbon blender, twin-shell blender, plowshare mixer, or equivalent.

No. 4

(Anionic, Light-Duty)

		Mixing Order
Bio Soft S-100	19.6	8
Sodium Carbonate (light, granular)	40.8	1
Sodium Bicarbonate	6.0	2

Steol CS-460	3.5	7
Snowlite II	69.8	3
Sodium Tripolyphosphate (light, granular)	8.0	4
Monosodium Phosphate (monohydrate)	6.0	5
Sodium Sesquicarbonate (crystals)	25.7	6
Sodium Sulfate	15.6	9
Water	1.0	8'
Sodium Silico Aluminate	4.0	10

Procedure:
Combine 1, 2, 3, 4, 5, and 6. Blend about 10 min. Atomize 7 into dry blend. Then mix 5–10 min. Combine 8 and 8'. Heat to 50 C and atomize about ½ of 8/8' into batch; then add 9 and continue mixing. Atomize balance of 8/8' into batch. Mix 30–45 min. Add 10, mix 15 min and pack out via 10 mesh sieve.

> Appearance: White, medium coarse granules.
> pH: 1% sol'n. = 9.78
>   0.1% sol'n. = 9.72
> Density (tamped): 0.80 g/ml

Equipment:
Ribbon blender, plowshare mixer, twin-shell blender, or equivalent.

No. 5

(Anionic, Light-Duty)

		Mixing Order
Bio Soft S-100	9.8	7
Sodium Tripolyphosphate (light, granular)	19.0	1
Sodium Sulfate	38.5	2
Sodium Chloride	9.0	3
Sodium Hydroxide (33% sol'n.)	3.5	8
Monosodium Phosphate (monohydrate)	5.0	9
Sodium Carbonate (light, granular)	5.5	4
Sodium Bicarbonate (powder)	5.5	5
3A Alcohol	0.7	6
Sodium Silico Aluminate	3.5	10

Procedure:
Combine 1, 2, 3, 4, and 5. Blend 5 min. Atomize 6 into dry blend. Continue mixing. Heat 7 to 50 C and atomize simultaneously with 8 into batch (requires two separate lines for atomizing). Add ½ of 10 when atomization is nearly completed. Continue mixing (ca. 5 min). Then add 9 and balance of 10. Mix until uniformly granulated and pack out via 10 mesh sieve.

Equipment:
Ribbon blender, twin-shell blender, rotating drum, or equivalent.

No. 6

(Anionic, Light-Duty)

		Mixing Order
Bio Soft S-100	9.8	8
Sodium Tripolyphosphate (light, granular)	20.0	1
Sodium Sesquicarbonate (crystals)	35.2	4
Sodium Carbonate (light, granular)	10.0	7
Monosodium Phosphate (monohydrate)	5.0	9
Sodium Sulfate	7.0	2
Sodium Chloride	7.0	3
Sodium Hydroxide (33% sol'n.)	3.5	5
Sodium Silico Aluminate	1.0	10
3A Alcohol	1.5	6

Procedure:
Combine 1, 2, 3, and 4; mix 5 min. Atomize 5 and then 6 into dry blend. Mix until uniform. Add 7 and mix 3-5 min. Heat 8 to 40-50 C and atomize slowly into batch. Mix about 10 min to complete reaction of 8 with alkalies. Add 9 and 10; mix 10 min and pack out via 10 mesh sieve.

Appearance: White; mixture of medium and fine granules.
pH: 1% sol'n. = 9.9
0.1% sol'n. = 9.8
Density (tamped): 0.77 g/ml

Equipment:
Ribbon blender, twin-shell blender, plowshare or auger mixer, rotating drum, or equivalent.

No. 7

(Anionic, Light-Duty)

		Mixing Order
Bio Soft S-100	9.8	8
Sodium Carbonate (light, granular)	19.6	1
Sodium Sesquicarbonate (crystals)	22.0	2
Monosodium Phosphate (monohydrate)	8.0	9
Sodium Tripolyphosphate (light, granular)	10.0	3
Sodium Silico Aluminate	2.0	10
Snowlite II	12.6	4
Sodium Chloride	7.5	5
Sodium Sulfate	7.5	6
Water	1.0	7

Procedure:
Combine 1, 2, 3, 4, 5, and 6; blend 5 min. Atomize 7 into dry blend; continue mixing. Heat 8 to ca. 50 C and atomize into batch. Mix 10–20 min. Add 9 and 10; mix 5 min and pack out via 10 mesh sieve.

Appearance: White, medium coarse granules.
pH: 1% sol'n. = 9.85
0.1% sol'n. = 9.7
Density (tamped): 0.85 g/ml

Equipment:
Ribbon blender, twin-shell blender, plowshare or auger mixer, rotating drum, or equivalent.

No. 8

(Anionic, Light-Duty)

		Mixing Order
Bio Soft S-100	9.0	6
Steol CS-460	2.0	5
Sodium Carbonate (light, granular)	20.0	1

Snowlite II	49.0	2
Sodium Tripolyphosphate (light, granular)	6.0	3
Monosodium Phosphate (monohydrate)	4.5	7
Sodium Sulfate	7.5	4
Sodium Silico Aluminate	2.0	8

Procedure:

Combine 1, 2, 3, and 4; blend 5 min. Atomize 5 into dry blend and continue mixing. Heat 6 to about 45-50 C and atomize into batch. At first signs of lumping, add 50% of 8. Continue mixing. Add 7 and mix 3-5 min. Finally, add balance of 8; mix 5-10 min and pack out via 10 mesh sieve.

> Appearance: White, medium coarse granules.
> pH: 1% sol'n. = 9.8
>   0.1% sol'n. = 9.7
> Density (tamped): 0.71 g/ml

Equipment:

Ribbon blender, twin-shell blender, plowshare or auger mixer, rotating drum, or equivalent.

No. 9

(Nonionic)

Sodium Dodecyl Benzene Sulfonate (40% flake)	25
Clindrol 200-L	3
Sodium Tripolyphosphate	32
Sodium Sesquicarbonate	40

Note:

The household liquid dishwashing detergent formula represents a very low cost product if prepared with the alkyl aryl sulfonic acid and caustic. The formula can be reduced in cost per gallon by increasing the water content to 70%.

No. 10

(Nonionic-Anionic)

		Mixing Order
Bio Soft S-100	6.0	7
Lathanol LAL Powder	2.0	9
Makon 12	2.0	12
Ninol AA-62 Extra	2.0	13
3A Alcohol	0.7	6
Sodium Sulfate	38.6	1
Sodium Chloride	7.5	2
Sodium Tripolyphosphate (light, granular)	19.5	3
Monosodium Phosphate (monohydrate)	5.0	10
Sodium Carbonate (light, granular)	5.5	4
Sodium Bicarbonate (powder)	5.5	5
Sodium Silico Aluminate	3.5	11
Sodium Hydroxide (33% sol'n.)	2.2	8

Procedure:

Combine 1, 2, 3, 4, and 5; mix 5 min and atomize 6 into blend. Heat 7 to 50 C. Atomize 7 and 8 simultaneously into blend. Add 9, 10, and 1/3 of 11, mixing 5 min after each addition. Combine 12 and 13 and heat to ca. 50-60 C. Atomize mixture into batch. Mix additional 5 min. Add balance of 11, mix to perfect granulation, and then pack out via 10 mesh sieve.

> Appearance: White; extremely uniform, medium-coarse granules.
> Tamped Density: 0.90 g/ml
> pH: 1% sol'n. = 9.1
>   0.1% sol'n. = 9.0

Equipment:

Ribbon blender, twin shell blender, rotating drum, or equivalent.

No. 11

(Nonionic-Anionic, Light-Duty)

		Mixing Order
Bio Soft EA-10	3.00	6

Bio Soft S-100	5.55	5
Steol CS-460	2.00	4
Sodium Carbonate (light, granular)	10.45	1
Stepan LMMEA	1.00	6'
Snowlite II	45.70	2
Sodium Silico Aluminate	1.00	7
Sodium Sulfate	31.30	3

Procedure:
Combine 1, 2, and 3; blend for 5 min. Atomize 4 into dry blend; mix 5 min. Heat 5 to 50 C and atomize into batch.* Combine 6 and 6'. Heat to dissolve 6' and atomize into batch. Add 7, mix 15 min, and pack out via 10 mesh sieve.

---
* Heating batch to ca. 40 C accelerates reaction of **Bio Soft S-100** with alkali and reduces mixing cycle.

Appearance: White, medium coarse granules.
pH: 1% sol'n. = 9.75
    0.1% sol'n. = 9.65
Density (tamped): 0.85 g/ml

Equipment:
Ribbon blender, twin-shell blender, plowshare or auger mixer, baffled rotating drum, or equivalent.

No. 12

(Nonionic-Anionic)

		Mixing Order
Makon 10	5.3	5
Sodium Sesquicarbonate (low density)	47.5	1
Bio Soft S-100	9.8	5'
Steol CS-460	3.5	4
Sodium Carbonate (light, granular)	18.4	2
Sodium Tripolyphosphate (light, granular)	8.0	3
Amidox L-2	1.8	6
Sodium Silico Aluminate	3.5	8
Cab-O-Sil EH-5	2.2	7

Procedure:
   Combine 1, 2, and 3; blend for 5 min. Atomize 4 into dry blend. (Pre-heating batch to 40 C will accelerate next step.) Combine 5 and 5′; heat to 40-50 C, and atomize into batch. Melt 6 and atomize into batch. Add 7 and 8, mix 10 min, and pack out via 10 mesh sieve.

   Tamped Density = 0.65 g/ml
   Product consists of white, nearly spherical granules.
   pH, 1% sol'n. = 9.9
       0.1% sol'n. = 9.9

Equipment:
   Auger, ribbon, or plowshare blender; twin-shell blender; rotating drum; or equivalent.

No. 13

(Nonionic-Anionic)

		Mixing Order
Makon 10	5.3	4
Snowlite II (sodium sesquicarbonate)	17.6	1
Bio Soft S-100	9.8	7
Steol CS-460	3.5	3
Sodium Carbonate (light, granular)	18.4	6
Sodium Tripolyphosphate (light, granular)	8.2	2
Amidox L-2	1.8	4′
Sodium Silico Aluminate	1.8	8
Cab-O-Sil EH-5	2.2	9
Snowlite II (sodium sesquicarbonate)	31.4	5

Procedure:
   Combine 1 and 2, blend 5 min, and atomize 3 into mixture. Combine 4 and 4′; heat to dissolve 4′, and atomize 50% of mixture into batch. Then add 5 and finish atomizing 4/4′. Charge 6. Heat 7 to 50 C and atomize into batch. Add 8, mix 5 min and charge 9. Mix 10 min and pack out via 10 mesh sieve.

   Tamped Density = 0.67 g/ml

Product consists of white, nearly-spherical granules.
pH: 1% sol'n. = 9.90
    0.1% sol'n. = 9.85

Equipment:
Twin-shell blender, rotating drum, plowshare mixer, ribbon blender, or equivalent.

No. 14
(Nonionic-Anionic, Light-Duty)

		Mixing Order
Nacconol 40DB	30.7	10
Sodium Tripolyphosphate (anhydrous)	13.8	1
Steol CS-460	7.4	8
Soda Ash	3.2	2
Borax (decahydrate, technical)	2.5	3
Ninol AA-62 Extra	3.7	9
Sodium Sesquicarbonate (L.D.)	8.1	4
Sodium Hexametaphosphate	6.1	5
Tetrasodium Pyrophosphate (anhydrous)	4.5	6
Sodium Sulfate	16.6	7
FD&C Green No. 3	q.s.	11
FD&C Yellow No. 5	q.s.	12
Cab-O-Sil M-5	3.4	14
Fragrance	q.s.	13

Procedure:
Combine ingredients 1-7; mix 15 min. Dissolve 9 in 8 and atomize into dry blend. Mix until homogeneous. Add 10 and mix 5 min.

Combine one part of 11 with three parts of 12; add dye mixture until desired shade is obtained. Finally add 13 and 14. Mix until fragrance is uniformly distributed. Screen through 12 mesh sieve if necessary and pack out.

Equipment:
Auger, plow-share, or twin-cone mixer; ribbon blender, roller mill or equivalent.

No. 15

(Nonionic-Anionic)

		Mixing Order
Ninol AA-62	3	3
Sodium Sesquicarbonate	42	1
Sodium Tripolyphosphate	30	2
Ultrawet SK	25	4

Procedure:

Thin amide by heating (perhaps add a little alcohol also) and spray onto the phosphate and sesquicarbonate while tumbling. The **Ultrawet SK** is added last and mixed in briefly.

Automatic Dishwashing Detergent

Formula No. 1

(Nonionic, Cream)

A	**Veegum® HS**	3.0
	CMC 7M	1.0
	Water	60.0
B	**Plurafac RA-43**	2.4
	Plurafac D-25	—
	Tetrapotassium Pyrophosphate	29.2
	Potassium Phosphate (tribasic)	4.4
	Potassium Chloride	—

Procedure:

Prepare A by adding **Veegum** and **CMC** slowly to the water, agitating continually until smooth. Add **Plurafac** to A with stirring. Add the salts with stirring. Heat is generated by the addition of the salts. Stir until batch is cool.

Packaging:

These formulas are fluid creams, and may be dispensed from plastic tubes or bottles, thus eliminating the need for handling dusty, irritating powders.

Directions for use:

Add one quarter to one half cup to the machine depending on the amount to be washed.

Comments:

This formula contains **Veegum HS** and **CMC** as the thickening agent to form a stable liquid cream. Potassium phosphates have been used as the soluble detergent builders and sequestering agents. **Plurafac RA-43** is a biodegradable, liquid, nonionic, low sudsing surfactant designed specifically for automatic dish washing machines. The use of a liquid cream eliminates the dusty condition associated with powders.

No. 2

(Nonionic, Powder)

		Mixing Order
Makon NF-5	6	3
Sodium Tripolyphosphate	47	1
Sodium Metasilicate (anhydrous)	47	2

Procedure:

Spray the **Makon NF-5** on the phosphate and silicate while they are being tumbled.

Note:

This high active formulation can be cut with 20–60% soda ash or inactive fillers.

No. 3

(Nonionic, Powder)

		Mixing Order
Makon NF-12	3.0	4
Sodium Metasilicate (pentahydrate)	41.0	1
Sodium Tripolyphosphate (light, granular)	41.6	2
Sequestrene NA-4	2.5	5

Sodium Hexametaphosphate	8.9	3
Sodium Dichloroisocyanurate (dihydrate)	3.0	6
Fragrance and Colorant	optional	

Procedure:
Combine 1, 2, and 3 in mixer; blend 5 min. Heat 4 to about 40 C and atomize into dry blend. Continue mixing for 5-10 min. Add 5 and 6; mix about 5 min. Add perfume and colorant as desired, mix until uniformly colored, and pack out via 10 mesh sieve.

Equipment:
Ribbon blender, twin-shell blender, plowshare or auger mixer, rotating drum, or equivalent.

No. 4

(Nonionic, Powder)

		Mixing Order
Makon NF-12	3.0	5
Sodium Metasilicate	32.0	1
Sodium Carbonate (light, granular)	25.3	3
Sodium Hexametaphosphate	9.7	4
Sequestrene NA-4	4.0	6
Sodium Dichloroisocyanurate (dihydrate)	3.0	7
Sodium Gluconate	2.0	8
Sodium Silico Aluminate	2.0	9

Procedure:
Combine 1, 2, 3, and 4 in mixer; blend 5 min. Heat 5 to about 40 C and atomize into dry blend; continue mixing 5-10 min. Then, add 6, 7, and 8; blend 5 min. Finally, add 9, mix 5-10 min, and pack out via 10 mesh sieve.

Equipment:
Ribbon blender, twin-shell blender, plowshare or auger mixer, rotating drum, or equivalent.

No. 5

(Nonionic, Powder)

		Mixing Order
Makon NF-12	3.0	3
Sodium Tripolyphosphate (anhydrous)	63.0	1
Sodium Silicate* (liquid)	31.0	2
Potassium Dichloroisocyanurate	1.3	6
Sodium Gluconate	1.3	7
Silicone L-45 (visc. = 50 cts.)	0.2	5
Fragrance (lemon/lime)	0.2	4
Colorant	q.s.	8

Procedure:
Charge 1 to blender and start mixing. Slowly atomize 2 onto phosphate. Next, spray 3, 4, and 5 in sequence onto phosphate/silicate. After addition of 5, mix 15–20 min. Finally, add 6, 7, and 8 in sequence mixing 5 min after each addition. When product is uniformly colored, pack out via 10 mesh sieve.

Equipment:
Twin-shell blender, rotating drum, plowshare mixer, ribbon blender, or equivalent.

---

*$Na_2Si_4O_9$, 40° Baume, ($Na_2O \cdot 4SiO_2$)

No. 6

(Anionic)

Nacconol 40F	2
Tetrasodium Pyrophosphate	22
Trisodium Phosphate (hydrate)	19
Sodium Metasilicate (pentahydrate)	25
Sodium Carbonate	32

No. 7

(Anionic, Low Foaming)

Water	147.3
Chlorinated Trisodium Phosphate	28.2
Potassium Hydroxide (45%)	28.2
Sodium Silicate (47%)	159.8
Tetra Potassium Pyrophosphate (60%)	548.3
ESI-Terge 330	28.2

Procedure:
Add in order mentioned and mix for 15 min.

No. 8

(Nonionic-Anionic, Liquid)

1. Water	11.7
2. **Igepal CO-730** (1% aqueous sol'n.)	0.5
3. **Gantrez AN-149**	1.0
4. Potassium Hydroxide (50%)	3.0
5. RU Sodium Silicate (47%)	17.0
6. **Heliogen® Blue N Supra** Paste (1%)	0.5
7. **Antarox BL-330**	3.0
8. Tetrapotassium Pyrophosphate (TKPP) (60%)	63.3

Procedure:
Heat water (80–85 C) and add ingredients 2 and 3 in order. Agitate for 45 min with moderate speed. Then slowly add ingredients 4-8 in order. Maintain temperature and continue agitation for 15 min after adding ingredients.

	No. 9	No. 10
	(Amphoteric, Liquid)	
Water	36.0	52.0
Potassium Hydroxide (45%)	9.0	7.0
Kasil #1	27.0	20.0

Tetrapotassium Pyrophosphate	14.0	10.5
Potassium Carbonate	12.0	9.0
Miranol® J2M Conc.	2.0	1.5

No. 11

(Inorganic–Nonionic)

Sodium Tripolyphosphate	30.0
Sulfosil® P-491	30.0
CDB-63	1.0
Plurafac RA43	2.0
Inert Fillers*	37.0

* Sodium sulfate, sodium chloride, or combination of both.

No. 12

(Inorganic–Nonionic)

Sodium Tripolyphosphate	30.0
Sodasil® P-598	25.0
CDB-63	1.0
Plurafac RA43	2.0
Inert Fillers*	42.0

* Sodium sulfate, sodium chloride, or combination of both.

Rinse Aid Additive

Formula	No. 1	No. 2	No. 3	No. 4
	(Nonionic, 70% Active)		*(Nonionic, 50% Active)*	
Antarox BL-330	74.0	74.0	53.0	53.0
Isopropanol	8.5	–	19.5	–
Urea	–	7.0	–	14.0
Water	17.5	19.0	27.5	33.0
Cloud Point	ca 55 C		ca 55 C	

Antarox BL-344	78.0	55.5
Water	22.0	44.5
Cloud Point	ca 62 C	ca 55 C

No. 5
(Nonionic–Anionic)

		Mixing Order
Makon NF-5	50	3
Stepanate X	15	2
Water	35	1

Procedure:
Combine the ingredients and mix.

High-Foam Laundry Detergent
(Anionic)

Nacconol 40F	52.5
Sodium Tripolyphosphate (anhydrous)	40.0
Carboxymethylcellulose	1.0
Sodium Metasilicate (pentahydrate)	6.0
Optical Brightener	0.5
Perfume	as desired

Phosphate-Free Laundry Detergent
(Nonionic)

Sodium Metasilicate (anhydrous)	10-20
Soda Ash (anhydrous)	10-30
Trisodium NTA	10-30
Borax	10-30
Sodium DBS (beads)	10-20
Alkasurf LA-9	2- 6

Alkamide CME	2- 4
Unisol DS-100	0.5- 1
Color and Perfume	q.s.

Low-Phosphate Laundry Detergent Powder

(Nonionic)

A	Sodium Tripolyphosphate (granular)	8
	Sodium Silicate	10
	Soda Ash (light)	11.5
	Sodium Carboxy Methyl Cellulose	2
	Sodium Sesquicarbonate	30
B	Brightener DS 100	0.5
	Perfume	q.s.
	Alkamide CME	3
	Alkasurf LA-16	5
C	Alkyl Benzene Sulfonate (40%, powder)	30

Procedure:

Blend ingredients designated A. While mixing, add B which contains the previously melted **Alkamide CME**. Only blend until uniform. Add C and mix 2-3 min.

Light-Duty Liquid Laundry Detergent

Formula	No. 1	No. 2	No. 3	No. 4	No. 5
			(Anionic)		
Total Active Matter (%w)	40	33	25	15	10
Neodol 25-3A (60% AM)	22.7	18.3	12.5	8.3	5.5
C_{12} LAS[8] (60% AM)	36.7	30	25	13.5	9.2
Lauric Diethanolamide[1]	5	4	2.5	1.9	1.2
Sodium Xylene Sulfonate[2]	5	8.5	–	3.0	3.0
Ethanol[3]	5.0	2.7	1.8	1.2	0.8
Sodium Sulfate[4]	2.6	2.2	0.8	0.6	0.4

Sodium Chloride	—	—	—	1.0	1.5
Urea	—	—	4.3	—	—
Water, Dye and Perfume	q.s.[5]	q.s.[5]	q.s.[5]	q.s.[5]	q.s.[5]
Visc. (cps @ 25 C)	200	200	205	220	125
Clear Point[6] (C)	+4	−5	0	−1	−1
Foam Performance[7]	128	109	81	62	47

---

[1] Such as **P & G Amide 72** or **Rechamide M3**.
[2] Such as **Ultra SXS** (40% AM).
[3] Derived from **Neodol 25-3A** except formulation when 1.8%w added.
[4] None added when oleum-derived LAS is used.
[5] Quantity sufficient to total 100 %w. Formalin preservative 0.1%w may be included.
[6] Temperature above which the solution is clear homogeneous liquid.
[7] Measured relative to the Shell SM-1 standard as 100.
[8] Such as **Sulframin 1260**.

No. 6

(Anionic)

Water	65.85
Tinopal 5 BM	0.25
Ultra® Sulfate AE-3	12.4
Sulframin® 60T	21.3
Perfume	0.2

No. 7

(Nonionic–Anionic)

		Mixing Order
Bio Soft N-300	40.0	2
Ninol 2012E	2.0	4
Steol CS-460	8.0	5
Stepanate X	5.0	3
Dowicide A	0.1	6
Sodium Chloride	q.s. up to 2.0 max.	7
Water	42.9	1
Colorant and Fragrance	q.s.	—

Procedure:
Combine 1, 2, and 3; mix until clear. Heat to 40 C. Add 4 and adjust pH to 7.0–7.5 with 86% phosphoric acid. Add 5 and 6; mix until clear. Readjust pH if necessary to 7.0–7.5. Add 7 to build up viscosity as desired.

No. 8

(Nonionic-Anionic)

Alkasurf LA-16	10–30
Alkasurf LA-Acid	5–15
Diethanolamine	4–10
Alkamide CDO	3– 6
Tinopal CBS-X	0.2– 0.5
Preservative	q.s.
Color and Perfume	q.s.
Water	to 100

No. 9

(Amphoteric, Sanitizing)

Miranol® C2M-SF Conc.	15.0
Sodium Tripolyphosphate	3.0
Trisodium Phosphate	3.0
Quaternary Ammonium Salt Germicide (50% active)	2.0
Organic Sequestering Agent (30%)	1.0
Water	76.0

Note:
Where high water hardness is encountered, it is suggested that 1% of an organic sequestering agent be used for each 100 ppm water hardness.

Heavy-Duty Laundry Detergents

Formula No. 1

(Nonionic, Phosphate-Free, Low-Foaming Powder)

		Mixing Order
Bio Soft EA-8	15.0	1
Brightsil H-20 $(SiO_2/Na_2O = 2/1)$	10.0	8
Sodium Carbonate (light, granular)	57.0	4
Sodium Bicarbonate	13.0	5
Sodium Silico Aluminate	2.0	6
Sodium Carboxymethylcellulose	2.0	7
Tinopal RBS-200	0.1	2
Tinopal AMS	0.9	3
Colorant and Fragrance	q.s.	—

Procedure:

Combine 1, 2, 3 and colorant; mix until uniform. Charge 4 to blender and administer solution (of 1-3 and colorant) by atomizing onto sodium carbonate. Mix until homogeneous. Add 5, 6, 7, and 8 in sequence mixing 2-3 min after each addition. Perfume batch as desired during the final mixing stage. Pack out via 10 mesh sieve.

Equipment:

Twin-shell mixer, rotating drum, plowshare blender, ribbon mixer, or equivalent.

	No. 2	No. 3	No. 4	No. 5
	(Nonionic, Phosphate)			
Neodol 25-7 (Neodol 25-9, Neodol 25-12)	10.0	10.0	10.0	10.0
STPP (low density)	40.0	40.0	40.0	40.0
Sodium Metasilicate Penta-hydrate[1]	5.0	5.0	5.0	—
Sodium Metasilicate Penta-hydrate[2]	—	—	—	5.0

CMC[3]	0.5	0.5	0.5	0.5
Na_2SO_4	34.5	29.5	33.5	34.5
Na_2CO_3 (anhydrous)	–	5.0	–	–
Fumed Silica[4]	–	–	1.0[5]	–
Water	10.0	10.0	10.0	10.0
Particle Distribution (%)				
> 40 mesh	62.7	66.7	85.6	71.3
> 40 > 100	34.1	31.1	14.3	24.4
< 100 mesh	3.2	2.2	0.1	4.3
Loose Density (g/cc)	0.645	0.715	0.614	0.664
Powder Flowability, % (DMSL[7]				
Funnel Method)	71.0	70.5	76.0	80.2
Time After Blending for				
Acceptable Flow (h)	1.75[6]	1.8[6]	0[6]	0.75[6]

[1] Metso-Granular.
[2] Metso 20.
[3] Hercules 7LT (95%)
[4] Cab-O-Sil 4-5
[5] Added after the water to speed up the powder aging or drying process.
[6] Prepared by making a nonionic STPP premix.
[7] Shell Detergent Market Support Laboratory.

Procedure:

The batch blending procedure that gave the best results in the laboratory with the phosphate containing formulation was the following, exemplified with **Neodol® 25-7**.

1. Prepare a **Neodol 25-7**/light density STPP concentrate by adding warm nonionic to the STPP over a 1-2 min time period while stirring.
2. Add the balance of the dry components and blend slowly for 1-2 min or sufficiently long enough to obtain a uniform product.
3. Add the water last with stirring and then continue to stir for 1-2 min or longer to break up any agglomerates that form.

As prepared, the blended powder is somewhat tacky and requires an aging or conditioning period before giving a flowability of 60% or greater, usually necessary for package filling. Using the premix or concentrate procedure, the aging period is approximately 1.75 h with **Metso Granular** (Formula No. 2) and 0.75 h with **Metso 20** (Formula No. 5). Combining all the dry components and then adding the nonionic and water in order resulted in a longer aging period of 2.25 h with both silicates. Partial replacement of Na_2SO_4 with Na_2CO_3 in Formula No. 3 also gave no reduction in aging time. However, the aging period can be reduced to one hour and zero hours, respectively, by the addition of 0.5 and 1.0% of **Cab-O-Sil H-5** after the addition of water (Formula No. 4). Adding 0.5% anhydrous Na_2CO_3 after the water resulted in only a marginal improvement in powder flowability. An attempt to leave the water out of the formulation gave a powder that was still wet and sticky after aging for 72 h.

Optical bleach (brighteners), perfume, and dye were not included in these blending experiments but can be added with an appropriate adjustment in the sodium sulfate content. The STPP should be a light density grade (available from Monsanto and FMC) and the sodium silicate similar to **Metso Granular** or **Metso 20** (sodium metasilicate pentahydrate) available from Philadelphia Quartz.

No. 6

(Nonionic)

Sodium Tripolyphosphate	10-40
Trisodium NTA	10-20
Sodium Metasilicate (anhydrous)	20-30
Sodium Sulfate (anhydrous)	10-20
Sodium CMC	2- 4
Alkasurf LA-9	6- 8
Alkamide HT-ME	0.5- 1
Unisol DS-100	2- 4
Color and Perfume	q.s.

No. 7

(Anionic)

Nacconol 40F	45.0
Tallow Soap	5.0
Sodium Tripolyphosphate (anhydrous)	41.0
Sodium Metasilicate (anhydrous)	7.5
Carboxymethylcellulose	1.0
Optical Brightener	0.5
Perfume	as desired

	No. 8	No. 9	No. 10	No. 11	No. 12	No. 13	No. 14	No. 15
				(Nonionic-Anionic, Nonphosphate)				
Neodol 25-7 (Neodol 25-9, Neodol 25-12	10	10	15	15	12	10	10	
Neodol 25-3S (active basis)	–	–	–	–	..	5		15
LAS (active basis)	..	–	–		–	..	5	–
Sodium Carbonate (anhydrous)	30	30	.	60	30	40	40	40
Sodium Sesquicarbonate	..	10	.	..	.	..	–	.
Sodium Citrate · 2H$_2$O	.	–	–	–	20	–	..	
Sodium Metasilicate · 5H$_2$O (1/1 SiO$_2$/Na$_2$O)	10	7	–	–	..	–	.	..
Sodium Silicate (2/1 SiO$_2$/Na$_2$O Ratio)	..	–	25	10	10	15	15	15
CMC	1	1	2	2	2	3	3	3
Sodium Sulfate	49	42	58	13	26	27	27	27

[1] Loose density, g/cc – 0.918
[2] Loose density, g/cc – 0.826
[3] Metso Granular.

Procedure:
The batch blending procedure that gave the best results in the laboratory with the low density granular sodium carbonate containing formulation was the following, exemplified with **Neodol 25-7**. A sodium carbonate pore volume of 30.6% is maximum.

1. Prepare a **Neodol 25-7**/sodium carbonate concentrate with a 25% nonionic concentration by adding the warm nonionic to the sodium carbonate over the 1-2 min period while stirring. At this nonionic loading the concentrate is still somewhat wet.
2. Age for 1 h, then blend in the balance of the dry components and blend slowly for 1-2 min or sufficiently long enough to obtain a uniform product.

[1] Loose density, g/cc – 0.918
[2] Loose density, g/cc – 0.826
[3] Metso Granular.

	No. 16	No. 17	No. 18	No. 19
	(Nonionic-Anionic, Liquid)			
Active Matter	50	40	30	40
Neodol 25-7	37.5	30.0	22.5	32.0
C_{12} LAS[1] (60% AM)	20.8	16.7	12.5	–
Neodol 25-3S (60% AM)	–	–	–	13.3
Triethanolamine (85% grade)	5.0	5.0	5.0	5.0
Ethanol SD-3A	6.0	5.5	5.0	5.0[2]
Potassium Chloride	1.0	1.0	2.0	4.0
Tinopal RBS (200%)[3]	0.025	0.025	0.025	0.025
Tinopal 5 BM[3]	0.1	0.1	0.1	0.1
Polar Brilliant Blue (RAWL 100%)[1]	0.01	0.01	0.01	0.01
Perfume–Tergescent No. 7[4]	0.10	0.10	0.10	0.10
Water	q.s.	q.s.	q.s.	q.s.
Visc. cps, 24 C (76 F)	125	131	108	143
Clear Point, C	8	−5	< −1	4
F	47	23	< 30	39
Cloud Point, C (5% sol'n.)	> 77	> 77	> 77	> 77
F	> 170	> 170	> 170	> 170
Pour Point, C	−15	−11	−6	–
F[6]	5	13	21	–
Flash Point, C	48	49	50	47
F	118	120	122	117
pH	10.5	10.4	10.5	9.2
Freeze-Thaw Test (3 cycles)	passes	passes	passes	passes
High Temp. Stability (49 C/120 F, 1 week)	passes	passes	passes	passes
Foaming Characteristic	passes	passes	passes	passes

[1] **Sulframin 1260.**
[2] In addition to that present in **Neodol 25-3S.**
[3] Optical brightener.
[4] Givauden Corp.
[5] Quantity sufficient to make 100% total.
[6] Pensky-Martens closed cup.

Procedure – Heavy-Duty Liquids:
    For the preparation of unbuilt, clear type formulations 16–19 the order

of addition is of importance to minimize viscosity resistance to mixing and to avoid possible gel formation. Effective stirring should be maintained during addition of all ingredients, and each ingredient should be in solution before the next is added. A blending temperature somewhat above ambient (e.g., 80-90 F) is recommended but not essential.

Put the water into the mixing vessel and add the KCl. Be sure that the amount of water added is corrected for the water content of the LAS slurry used. Add the ethanol and triethanolamine. Add the sodium alkylbenzene sulfonate (LAS) slurry. Add the Neodol® 25-7. Be sure to have efficient stirring, and, if feasible, add the Neodol 25-7 near the vortex of the stirrer. Add the fluorescent whitening agent with rapid stirring. Add perfume and dye as needed to give the desired odor and color.

The brighteners are the slowest dissolving of all the ingredients. Attempts were made to predissolve the Tinopals in the Neodol 25-7, the ethanol, the TEA, and a mixture of the three. In no case did the Tinopal mixture dissolve more rapidly or completely than it did in the total formulation.

No. 20

(Nonionic-Anionic, Liquid)

Water	20.0
Ultra® SXS Liquid	25.0
Triethanolamine	1.1
Sulframin® 1260 Slurry	17.9
Tergitol 15-S7	36.0
Fluorescent Whitener Hiltamine CWD	0.25

No. 21

(Nonionic-Anionic, Nonphosphate, Controlled Foam Powder)

		Mixing Order
Nacconol 90F	13.0	6
Bio Soft EA-10	4.0	1
Brightsil H-20 ($SiO_2/Na_2O = 2/1$)	12.0	9
Sodium Carbonate (light, granular)	53.9	5

Sodium Bicarbonate	12.0	7
Sodium Silico Aluminate	2.0	8
Sodium Carboxymethylcellulose	2.4	10
Tinopal AMS	0.5	2
Tinopal 5BM	0.1	3
Tinopal RBS-200	0.1	4
Colorant and Perfume	q.s.	

Procedure:
  Combine 1, 2, 3, 4, and colorant; mix until homogeneous. Charge 5 and 6 to blender and mix 5 min. Then atomize solution of 1-5 and colorant onto dry mixture. Mix 5 min. Add 7, 8, 9, and 10 in sequence mixing 2-3 min after each addition. Finally, atomize perfume into batch; mix 5 min and pack out via 10 mesh sieve.

Equipment:
  Twin-shell mixer, rotating drum, plowshare blender, or equivalent.

No. 22

(Nonionic-Anionic, Phosphate, Moderate Foaming Powder)

		Mixing Order
Nacconol 90F	14.0	6
Bio Soft EA-8	6.0	1
Sodium Tripolyphosphate (light, granular)	35.0	3
Metso Pentabead 20 (pentahydrate)	9.0	4
Snowsil	20.0	5
Puffed Borax (12-14#/ft^3)	11.2	9
Sodium Carboxymethylcellulose	2.0	8
Tinopal RBS-200	0.8	2
Sodium Silico Aluminate	2.0	7
Colorant and Perfume	q.s.	—

Procedure:
  Combine 1, 2, and colorant. Mix until homogeneous. Charge 3, 4, 5, and 6 to blender. Mix 5 min and atomize solution of 1, 2, and colorant onto this mixture. Add 7, 8, and 9 in sequence mixing 2-3 min after each

addition. Finally, atomize perfume into batch; mix 5-10 min and pack out via 10 mesh sieve.

Equipment:
    Rotating drum, twin-shell mixer, ribbon blender, plowshare mixer, or equivalent.

No. 23

(Anionic-Inorganic, Nonphosphate)

Sulframin® 40RA Beads	35
Sodasil P-598	50
Sodium Sulfate	14.2
Carboxymethylcellulose	0.5
Fluorescent Whitening Agent	0.2
Perfume	0.1

Note:
    If the density is not of major importance, replace **40RA** with **Sulframin® 40 Flake** or **Granular**. This will substantially reduce the cost.

No. 24

(Anionic-Inorganic)

Sulframin® 40RA Beads	45.0
Sodium Tripolyphosphate	22.0
Sulfosil® P-491	32.2
Carboxymethylcellulose	0.5
Fluorescent Whitening Agent	0.2
Perfume	0.1

Note:
    This will yield a bulk low density product. If the density is not of major importance, replace **40RA** with **Sulframin® 40 Flake** or **Granular**. This will substantially reduce the cost.

Pump-Spray Laundry Prewash

Formula No. 1

(Nonionic-Anionic)

Neodol® 25-3 (Neodol 23-3)	10
Neodol 23-6.5	10
Shell Sol 71[1]	20
Isopropyl Alcohol	12
Triethanolamine Oleate[2]	3.5
Water	q.s.[3]

[1] Isoparaffinic solvent, bp 179.4–204.4 C.
[2] Can be prepared in situ from triethanolamine and oleic acid.
[3] Quantity sufficient to make 100% total.

No. 2

(Nonionic-Anionic)

Neodol 25-3 (Neodol 23-3)	22
Shell Sol® 71[1]	18
Sodium Xylene Sulfonate (40% AM)[2]	30
Triethanolamine Oleate[3]	2.5
Water	q.s.[4]

[1] Isoparaffinic solvent, bp 179.4–204.4 C.
[2] Ultra SXS.
[3] Can be prepared in situ from triethanolamine and oleic acid.
[4] Quantity sufficient to make 100% total.

Laundry Softener

Formula No. 1

Conco Softener CC-75	9.0
Tinopal BHS	0.2
Perfume	q.s.
Coloring	q.s.
Water	q.s. to 100

No. 2

(Amphoteric)

Miranol® SHD Conc.	3.500
Dyestuff	0.003
Perfume Oil	0.050
Phosphoric Acid, 85% (to reduce the pH to 3.5)	0.040
Water	96.407

Note:

Stable dispersions of **Miranol® SHD Conc.** in water should be prepared by adding the water at 70-80 C.

	No. 3	No. 4	No. 5	No. 6
		(Cationic)		
Alkaquat T	4.0	5.4	6.7	8.0
Water	96.0	94.6	93.3	92.0
Color and Perfume	q.s.	q.s.	q.s.	q.s.

Procedure:

Charge the water to a suitable mixing vessel and preheat to 110-120 F. With gentle agitation, add the **Alkaquat T** previously warmed to room temperature if necessary. Continue agitation until a smooth dispersion is obtained (about 15-20 min). Add color and perfume during this period. Adjust pH to 4.5-5.5. The mixture is now cooled to bottling temperature.

Directions for use:

8-10 lb wash load	4-5 oz	3-4 oz	2-3 oz	1½-2 oz

Note:

Excellent pink shades can be achieved from Rhodamine B Extra (B.A.S.F.) and good blue shades can be obtained from Ceres Blue GN (Bayer).

Detergent Reodorant

(Nonionic)

Pine Oil	8.50
Volpo T15	17.50
Water (deionized)	73.80

Formalin	0.20
Color	q.s.

Procedure:
Blend pine oil and **Volpo** by warming. Add hot water and stir.

Fine Fabric Detergent

(Anionic)

Igepon T-73	30.0
Sodium Hexametaphosphate	35.0
Sodium Bicarbonate	34.2
Blancophor® FFG Brightener	0.3
PVP K-30 (polyvinylpyrrolidone)	0.5

Synthetic Cleaner

(Nonionic)

Water	89.00
Trisodium Phosphate	0.80
Sodium Tripolyphosphate	1.60
Potassium Hydroxide (90%)	0.30
Tall Oil Fatty Acid	3.30
ESI-Terge B-15	5.00

Procedure:
Add in order listed with adequate agitation. Allow all powders to dissolve before adding tall oil fatty acid and **ESI-Terge B-15**. Agitate until clear.

Liquid Nylon Wash and Brightener

FORMULA No. 1

(Nonionic-Anionic)

		Mixing Order
Bio Soft D-60	36.0	2
Steol CS-460	10.0	4
Makon 10	10.0	3

Ninol AA-62	6.0	5
Water	37.5	1
Blancophor® FFG (nylon substantive optical brightener)	0.5	6

Procedure:
   Combine ingredients and stir until uniform. Heat may be necessary to facilitate solution or dispersion of the **Blancophor.**

No. 2

(Anionic, Liquid)

		Mixing Order
Bio Soft N-300	20.0	2
Stepan HDA-7	22.3	3
Tinopal RBS-200	0.1	4
Methyl Paraben	0.1	5
Water	57.5	1
Fragrance	q.s.	—

Procedure:
   Combine ingredients in the order listed mixing 2-3 min after each addition. Heat to 40 C if desired to hasten dissolution of 4 and 5.

No. 3

(Anionic-Nonionic)

		Mixing Order
Bio Soft N-300	25.00	4
Steol CS-460	22.5	5
Ninol AA-62 Extra	4.17	6
Stepanate X	10.4	3
Tetra Potassium Pyrophosphate	3.33	2
Water (deionized)	34.6	1

Procedure:
Dissolve the TKPP in the water and add the other ingredients in the order indicated with good agitation. The **Ninol AA-62 Extra** should be melted before adding.

No. 4

		Mixing Order
Bio Terge M-300	30.00	5
DEA Cocoate (soap)	23.00	4
Ottasept – Extra	0.20	3
Tinopal RBS (200%)	0.02	2
Water	46.78	1

Procedure:
Blend into the water the **Tinopal, Ottasept,** DEA Cocoate and **Bio Terge M-300.** Heat to 70 C with good mixing. Cool to room temperature with mixing.

Cold Water Wool Detergent

Formula No. 1

(Anionic)

		Mixing Order
Stepanol 317	30.00	4
Tetrapotassium Pyrophosphate	4.00	3
Stepanate X	10.00	5
Tinopal RBS-200	0.02	2
Water	55.98	1

Procedure:
Dissolve phosphate and **Tinopal RBS-200** in water and **Stepanate X.** Add **Stepanol 317** and stir until uniform.

Hand Wash Cold Water Liquid Detergents

	Formula No. 1	No. 2
		(Nonionic–Anionic)
Neodol® 25-7	–	10
Neodol 25-3S (60% AM)	25	0
Lauric Diethanolamide[1]	5	5
Cocoamido Betaine (30% AM)[2]	–	15
Water, Brightness, Perfume, etc.	q.s.	q.s.[3]

[1] Nitrene 6180, or equivalent.
[2] Monateric CAB, or equivalent.
[3] Quantity sufficient to make 100% total.

Diaper-Rinse Tablets
(Cationic)

A Hyamine 1622	18.0
Starch	47.4
Milk Sugar	26.6
Sodium Bicarbonate	9.0
B Alcohol	6.0
Water	90.0

Procedure:
Mix A; mix B. Add A to B stirring until uniform.

Diaper Rinse
(Nonionic)

Cetab	16.8
Water	73.2
Triton X-100	10.0

Procedure:
Dissolve the **Cetab** in the water, with a little heat, if necessary. Add the **Triton** and stir until the solution is homogeneous.

Use 1 teaspoonful (about 8 cc) of the antiseptic per gallon of warm water.

Scouring Cleanser

Formula No. 1

(Anionic, Powder)

Nacconol 40F	15
Ground Lumestone or **Super-Cel**	50
Sodium Tripolyphosphate (anhydrous)	15
Bentonite	10
Sodium Soap Powder	10

No. 2

(Anionic)

Sulframin® 40RA Beads	8.5
Sodium Tripolyphosphate	3.75
Trisodium Phosphate (anhydrous)	2.0
CDB-69	0.75
Pumice or Silica Flour (a fine particle size abrasive)	85.0

No. 3

(Anionic)

		Mixing Order
Nacconol 40F	3.0	5
Trisodium Phosphate	5.0	3
Sodium Tripolyphosphate	5.0	2
Silica Flour	80.0	1
Sodium Perborate (monohydrate)	7.0	4
Perfume	q.s.	

Procedure:

Combine 1, 2, 3, and 4 in blender; mix until homogeneous. Add 5 and mix 15 min. Atomize perfume into blend and mix until fragrance is uniformly distributed. Pack out via 20 mesh screen.

Equipment:
  Twin-shell blender, plowshare mixer, auger mixer, ribbon blender, rotating drum, or equivalent.

No. 4

(Anionic, Powder)

		Mixing Order
Nacconol 40F	3.0	8
Sodium Hexametaphosphate	5.0	7
Trisodium Phosphate (anhydrous)	5.0	6
Ground Limestone	56.7	1
Bentonite	10.0	2
Super Floss	15.0	3
Trisodium Phosphate (chlorinated)	5.0	4
Sodium Metasilicate (anhydrous)	0.3	5
Colorant and Perfume	q.s.	9

Procedure:
  Combine 1, 2, 3, 4, 5, 6, and 7 in blender; mix until uniform. Add 8 and mix 15 min. Add colorant as desired and mix to uniform shade. Spray or atomize perfume into blend, mix until fragrance is uniformly distributed, and pack out via 20 mesh screen.

Equipment:
  Twin-shell blender, ribbon blender, auger or plowshare mixer, rotating drum.

No. 5

(Anionic, Powder)

		Mixing Order
Nacconol 40F	2.3	8
Sodium Sesquicarbonate	5.2	6
Trisodium Phosphate (anhydrous)	5.5	5

Ground Limestone	53.5	1
Bentonite (325 mesh)	10.0	2
Super Floss	20.0	3
Potassium Dichloroisocyanurate	3.0	4
Sodium Metasilicate (anhydrous)	0.5	7
Colorand and Perfume	q.s.	9

Procedure:

Combine 1, 2, 3, 4, 5, 6, and 7 in blender and mix until uniform. Add 8 and mix 15 min. Add colorant as desired and mix until uniform. Spray or atomize perfume into mixture, mix until fragrance is uniformly distributed, and pack out via 20 mesh screen.

Equipment:

Twin-shell blender, ribbon blender, auger or plowshare mixer, rotating drum, or equivalent.

No. 6

(Nonionic, Liquid)

A	**Veegum®**	4.3
	Water	60.7
B	Calcium Carbonate (100 mesh)	30.0
C	**Plurafac D-25**	5.0
D	Perfume	q.s.
	Preservative	q.s.

Procedure:

Add the **Veegum** to the water slowly, agitating continually until smooth. Add the abrasive with stirring. Add the **Plurafac** and mix until smooth and uniform.

Packaging:

This formula is a creamy liquid and can be dispensed from a plastic squeeze bottle, thus eliminating the need for harsh dusty powders.

Comments:

In this formula **Veegum** is a suspending agent for the abrasive, and a thickening agent which forms a stable liquid cream. **Plurafac D-25** is a nonionic, biodegradable surfactant which supplies detergency and leaves no dulling film. Containing no bleaching agent, this composition is easily perfumed, is gentle to hands, and is pleasant to use.

Directions for use:
Apply a small amount to surface to be cleaned. Wipe with a damp sponge and rinse with water.

No. 7

(Nonionic, Foam)

		% in Aerosol
Concentrate:		90.0
Cab-O-Sil	3.5	
Triton X-100	4.0	
Butyl Cellosolve	7.5	
Mapleton 325 (Mesh Supersil)	10.0	
Water	75.0	
Propellent:		10.0
Isotron 12	40.0	
Isotron 114	60.0	

Package:
Lined tinplate container with a foam-type valve.

Procedure:
Dissolve the **Triton X-100** and butyl cellosolve in the water. Add the abrasive, followed by the **Cab-O-Sil**, with enough agitation to insure even dispersion. Should be pressure filled.

Directions for use:
Shake well before use. Foam a small amount on surface to be cleaned. Rub with a moistened cloth or sponge to desired luster.

Warning:
Contents under pressure. Do not puncture. Exposure to heat or prolonged exposure to sun may cause bursting. Do not throw into fire or incinerator. Keep from children.

No. 8

A	Veegum® HS	1.50
	Water	62.90
B	Calcium Carbonate (100 mesh)	21.00
	Super Floss	9.00
C	Sodium Lauryl Sulfate	0.35
D	Tetrapotassium Pyrophosphate	3.00
	Potassium Phosphate (tribasic)	1.50
E	Calcium Hypochlorite (70% active)	0.75

Procedure:
Add the **Veegum HS** to the water slowly, agitating continually until smooth. Add B to A and mix until smooth. Add C, D, and E in order, mixing after each until uniform.

Packaging:
This formula is a fluid cream and may best be dispensed from a plastic squeeze bottle, thus eliminating the need for harsh, dusty powders.

Comments:
In this formula, **Veegum HS** serves as a suspending aid for the mild abrasive system, and advantage is taken of its superior electrolyte stability to form a stable liquid cream. The phosphates are used as detergent builders and as buffering agents to maintain the composition pH at approximately 11. In this alkaline region the sodium lauryl sulfate and the hypochlorite bleach form a stable and effective cleaning system.

The use of a liquid cream offers the advantages of mildness in appearance and use as well as greater ease and effectiveness on vertical surfaces when compared to powders. In addition, compositions of this type are easily wiped or rinsed from hard surfaces, leaving no gritty film or residue.

Beverage Glass Cleaner

Formula No. 1

(Nonionic-Anionic)

		Mixing Order
Bio Soft D-60	3	1
Ninol AA-62 Extra	2	2
Water	95	3

Procedure:
Combine the ingredients and mix with heating (60 C) until **Ninol AA-62 Extra** goes into solution.

No. 2

(Liquid)

A	**Veegum**	1.5
	Water	67.5
B	Sorbitan Monolaurate	2.0
	Polysorbate 20	2.0
	Ammonia	2.0
	Deodorized Kerosene	15.0
	Abrasive	10.0
C	Preservative	q.s.

Procedure:
Add the **Veegum** to the water slowly, agitating continually until smooth. Add the components in B in the order listed, with mixing after each addition. Then add C.

Aerosol package: Concentrate 80%; Propellant 12 20%.

Oven Cleaner

Formula No. 1

(Anionic)

Potassium Hydroxide (45%)	85.0
ESI-Terge 330	15.0

Procedure:
Add as listed.

Active ingredients – 53.0%
Visc.: 3500 cps LV #1 Spindle @ 60 rpm; to increase viscosity to about 4500 cps increase **ESI-Terge** to 20% and decrease potassium hydroxide to 80%.

Warning:
Proper protective clothing should be used when handling potassium hydroxide. Heat is given off while reacting.

No. 2

(Nonionic, Aerosol)

A **Veegum T**		2.9
Water		86.4
B **Pluronic F-127**		4.3
C **Dow Corning 200 Fluid** (60,000 cs)		6.4

Procedure:
Add the **Veegum T** to the water slowly, agitating continually until smooth. Add B to A with agitation. Add C, mixing until uniform.

Aerosol package:
Concentrate 70%; propellant 12 30%.

Directions for use:
Spray evenly on clean unheated oven walls from a distance of about 10 in. After cooking allow oven to cool, then wipe off film and grease splatter with damp sponge.

No. 3

(Nonionic, Aerosol)

A **Veegum T**	2
Water	32
B Ammonium Hydroxide	8

1,1,1-Trichloroethane	24
Ethanol	10
Tergitol NPX	24

Procedure:
Add the **Veegum** to the water slowly, agitating continually until smooth. Add B to A and mix until uniform. Package in aerosol dispenser.

Aerosol package:
Concentrate 75%; Propellant 12/114 25%.

Directions for use:
Oven should be cool and empty. Do not use near flame. Contains alcohol. With can held upright, spray oven surface from a distance of 9-12 in. Allow foam to work for 10-20 min. Wipe clean with wet cloth or sponge.

No. 4

Olein (distilled)	40
Stearin	10

Mix warm.

Spindle Oil	40
Tetralin	9
Ammonia (sp. g. 0.91)	1
Emery, Pumice, or Tripoli sufficient to make pasty	

Procedure:
Dissolve stearin in olein by heating. Then mix the other items and stir both together until uniform.

No. 5

(Nonionic-Anionic, Foam-Spray)

Water	70.00
ESI-Terge 330	5.00
ESI-Terge 320	2.50
ESI-Terge S-10	2.50
Potassium Hydroxide (KOH) 90%	20.00

Procedure:
    Add water and **ESI-Terges** to a suitable tank (preferably a water jacketed kettle). Add potassium hydroxide (KOH) very slowly so temperature does not exceed 50 C.

Solids:	24-26%
Active:	26-28%
pH:	14
Free KOH:	14-17%
Visc.:	10 cps max LV #2 @ 60 rpm

Note:
    This cleaner forms a foam when used with proper foam trigger sprayer.

Warning:
    Proper protective clothing should be used when handling potassium hydroxide (KOH).

No. 6

(Amphoteric, Heavy-Duty)

A	**Veegum T**	0.75
	Kelzan	0.25
	Water	72.00
B	**Antaron FC-34**	2.00
C	Sodium Hydroxide	10.00
	Water	15.00

Procedure:
    Dry blend **Veegum T** and **Kelzan**. Add to the water slowly, agitating continually until smooth. Add B and stir slowly until uniform. Dissolve the sodium hydroxide in water. Add C slowly to the other components. Mix until uniform.

Aerosol package:
    Concentrate 80%; Propellant 12/114, 30/70 20%.

Directions for use:
    Use in either warm or cold oven. With can held upright, spray oven sur-

face from a distance of 9-12 in. Allow foam to work for 10-20 min. Wipe clean with wet cloth or sponge.

No. 7

(Amphoteric, Pump-Spray)

A	Veegum® T	1.2
	Kelzan	0.4
	Water	78.4
B	Sodium Hydroxide Sol'n. (40%)	10.0
C	Monateric Cy Na 50	10.0

Procedure:
Add the dry blend of **Veegum T** and **Kelzan** to the water slowly, agitating continually until smooth. Add B and C in order, mixing after each addition until smooth and uniform.

Packaging:
This formula is a fluid gel suitable for pump spray dispensing.

Directions for use:
Use in either warm or cold oven. Spray oven surface from a distance of 9-12 in. Allow foam to work for 10-20 min. Wipe clean with wet sponge or cloth.

Comments:
The use of **Veegum T** in this formula provides a stable fluid product which, when dispensed through a pump sprayer, shows good adhesion with no running on vertical surfaces. The amphoteric surfactant effectively cuts through oven soils, allowing the use of a relatively low level of caustic. Therefore, the product will be safer and less irritating to the user.

No. 8

(Caustic, Aerosol)

Volpo NP9	1.00
Volpo CO5	5.00
Methofas PM20	0.50
Caustic Soda Sol'n. (aqueous)	86.00
Propellent 12/114 (40:60)	7.50

Procedure:

The **Methofas** is slowly added to approximately one third of the water with constant stirring. The caustic is dissolved in the remaining water along with the **Volpos**. Finally, add the **Methofas** dispersion with stirring. Fill off and pack.

Packaging:

Aerosol packages containing strong caustic materials can give rise to corrosion problems. While the nature of this corrosion is not fully understood, it is believed that the oxygen content of the final pack is a contributory factor. To reduce the oxygen content of the pack, aeration of the product during manufacture must be avoided and the containers should be purged before filling with the propellant.

No. 9

(Nonaerosol)

Veegum T	0.75
Kelzan	0.25
Crodafos T5 Acid	2.00
Sodium Hydroxide	10.00
Water (deionized)	87.00

Procedure:

Ensure that the thickening agents **Veegum** and **Kelzan** are fully dispersed into the water prior to the addition of the **Crodafos** and caustic soda.

Warning:

Due to the corrosive effect described above, all formulations of this type should be thoroughly tested for effects on the containers before marketing. Individual country legislation concerning the packaging of hazardous compounds should be consulted.

Refrigerator Cleaner

Gloquat C	1
Sodium Bicarbonate	3
Water	96

Procedure:
Mix all ingredients together until uniform.

Note:
This solution incorporates a quaternary ammonium compound, making it a germicidal, antistatic cleaner.

Chapter III

POLISHES AND WAXES

Car Polish

Formula No. 1

(Anionic, Liquid Emulsion)

Syncrowax ERL	4.00
Silicone Fluid (350 cs)	2.00
White Spirit	15.00
Crodolene	3.00
Water (deionized)	71.00
Morpholine	1.00
Snowfloss	5.00
Formalin	0.20

Procedure:

Melt wax at 100 C, add oleic acid and silicone and run in solvent. Heat water to 90 C, add morpholine and run into oil phase with vigorous stirring. Homogenize, add **Snowfloss** and formalin at 45 C.

No. 2

(Anionic, Paste)

Syncrowax HGL	40.00
Polyethylene AC 629	10.00
Rocsol Wax COA	5.00
Paraffin Wax 140/145	45.00
Silicone Fluid (1000 cs)	5.00
White Spirit	325.00
Snowfloss	12.00

Procedure:
Melt waxes (110–120 C). Add silicone fluid followed by solvent, Silverson to 45 C. Add **Snowfloss**, cool and fill.

No. 3
(Anionic, Soft Paste)

Syncrowax HGL	40.00
Rocsol Wax COB	20.00
Paraffin Wax 140/145	40.00
Silicone Fluid (1000 cs)	2.00
Silicone Fluid (300 cs)	8.00
White Spirit	300.00
Snowfloss	12.00

Procedure:
Melt waxes at 100 C, add silicone fluid followed by solvent. Silverson to 45 C. Add **Snowfloss**, cool and fill.

No. 4
(Anionic, Abrasive)

Syncrowax ERL	4.00
Silicone Fluid (500 cs)	2.00
White Spirit or Distillate	15.00
Crill 2	1.50
Crillet 2	1.50
Neuberg Chalk	5.00
Water	71.00

No. 5
(Anionic)

A **SF96®** (100)	3.50
Wax S	2.60
Oleic Acid	1.50
Mineral Spirits	5.00

B Morpholine	1.40
Carbopol 934	0.17
Triethanolamine	0.17
Water	85.66

Procedure:

Heat A to 90 C (194 F). Heat B to 85 C (185 F). Add A to B with high-shear agitation. When homogeneous, continue mixing with mild agitation until cool.

Application:

Clean car surface before application. Pour a small amount on a dry, soft absorbent cloth and apply to a small section of the surface (2' X 2'). Wipe to a shine using a dry section of the cloth for the final shine.

Warning:

When solvents are used, as described above, proper safety precautions must be observed. All solvents must be considered toxic and should be used only in well ventilated areas. Prolonged exposure to solvent vapors must be avoided. If flammable solvents are used, storage, mixing, and use must be in areas away from open flames or other sources of ignition. The selection of any solvent, particularly chlorinated hydrocarbon solvents, will require consideration of applicable OSHA, EPA and other federal, state and local regulations.

No. 6

(Nonionic, Emulsion)

A **Viscasil®** (10,000)	4.0
C-36 Wax	3.0
Arlacel C	2.0
Mineral Spirits	17.0
Kerosene	8.0
B Water	56.0
C **Kaopolite SFO**	10.0

Procedure:

Melt **C-36** wax in Part A and add remaining ingredients of Part A preheated to 70 C (167 F). Mix well. Heat Part B water to 95 C (203 F) and add slowly to Part A with rapid high-shear agitation. Add Part C and blend until uniform. Continue mild agitation until cool.

Application:
Apply a thin, even coat with a damp cloth using a circular motion. Allow to dry. Remove white haze with a clean, dry cloth. Turn cloth frequently to obtain maximum ease of rubout.

Warning:
When solvents are used as described above, proper safety precautions must be observed. All solvents must be considered toxic and should be used only in well ventilated areas. Prolonged exposure to solvent vapors must be avoided. If flammable solvents are used, storage, mixing, and use must be in areas away from open flames or other sources of ignition. The selection of any solvent, particularly chlorinated hydrocarbon solvents, will require consideration of applicable OSHA, EPA, and other federal, state, and local regulations.

No. 7

(Nonionic)

A	SF 96® (1000)	5.7
	Viscasil® (10,000)	1.5
	Wax S	8.0
	Oleic Acid	1.0
	Mineral Spirits	10.0
B	Morpholine	0.8
C	Water	44.0
	Kaopolite SF	6.0
	Snow Floss	6.0
D	Mineral Spirits	7.0
	Kerosene	10.0

Procedure:
Heat components of Part A to 90 C (194 F) and hold at this temperature until all of the wax is melted. Add Part B slowly. Heat water to 90 C (194 F) and add abrasives to form a uniform slurry. Add Part C to A–B

and agitate until uniform. Add Part D slowly and continue mixing until wax starts to thicken. Pour into chilled containers. Circulating cool air across the container and paste surface will aid solidification. The pH of the product should be neutral or slightly acidic.

Warning:

When solvents are used as described above, proper safety precautions must be observed. All solvents must be considered toxic and should be used only in well ventilated areas. Prolonged exposure to solvent vapors must be avoided. If flammable solvents are used, storage, mixing, and use must be in areas away from open flames or other sources of ignition. The selection of any solvent, particularly chlorinated hydrocarbon solvents, will require consideration of applicable OSHA, EPA, and other federal, state, and local regulations.

No. 8

(Nonionic, Solvent-Based)

White Spirit	97.50
Silicone Fluid (350 cs)	1.50
Syncrowax HGL	1.00

Procedure:

Melt wax at 100 C, add silicone and white spirit, producing a hazy liquid product for quick application and removal.

No. 9

(Nonionic, Water-Based)

Carnauba Wax	0.40
Syncrowax ERL	0.90
Silicone Fluid (300 cs)	1.00
Armasol H	16.00
Crill 1	0.50
Volpo NP9	0.40
Water (deionized)	60.80
Propellant 12	20.00

Procedure:

The waxes and silicone are dissolved separately into the **Armasol H** at 80 C. **Crill 1** and **Volpo NP9** are dissolved in water and the solution heated to 80 C. The two solutions at 80 C are then mixed vigorously. The resulting emulsion is then allowed to cool to ambient temperature while continuing to stir. Finally the **Propellant 12** is pressure filled into cans containing the emulsion.

No. 10

(Nonionic, Abrasive)

Rocsol Wax CGO	4.00
Stearyl Alcohol	2.00
Volpo CO20	2.00
Silicone Fluid (800 cs)	2.00
Bentonite	7.00
White Mineral Spirits	20.00
Water	63.00

Procedure:

The wax, stearyl alcohol and **Volpo CO20** are melted. The mineral spirits containing the silicone are added, followed by bentonite. Finally warm water is run into the mixture, using rapid agitation. Cool while stirring.

Note:

Less mineral spirits or bentonite will result in thinner creams.

No. 11

(Nonionic-Anionic, Emulsion)

A	**SF96®** (1000)	5.00
	Wax S	2.00
	Oleic Acid	2.00
	Mineral Spirits	15.00
	Kerosene	15.00

B Morpholine	1.00
Carbopol 934	0.075
Triethanolamine	0.075
Water	49.85
C **Snow Floss**	10.00

Procedure:

Heat components of Part A to 90 C (194 F). In Part B disperse the **Carbopol 934** in water and when completely dissolved, add the triethanolamine. Heat mixture to 90 C (194 F). Then add the morpholine. Add Part A to Part B with rapid high-shear agitation until a uniform mixture is obtained. Cool to 40–45 C (104–113 F). Add Part C slowly and continue agitation until a uniform blend is obtained.

Aerosol packaging:

This formula can be either liquid packaged or used in a foam-type aerosol package. If the latter is desired, use the following procedure:

1. Check the viscosity of the material with a Brookfield viscometer, Model RVF-7, Spindle No. 4 for 15 s at 20° rpm. The viscosity should be between 3400–4000 cps @ 25 C.
2. Aerosol Loading   Parts by Weight
   Concentrate   94.5
   Isobutane   5.5

Application:

Remove excess dirt by washing. Apply polish to a clean, soft cloth and apply to the car surface using a circular motion. Allow the polish to dry to a white haze and then buff to a high gloss with a clean, dry cloth.

Warning:

When solvents are used as described above, proper safety precautions must be observed. All solvents must be considered toxic and should be used only in well ventilated areas. Prolonged exposure to solvent vapors must be avoided. If flammable solvents are used, storage, mixing, and use must be in areas away from open flames or other sources of ignition. The selection of any solvent, particularly chlorinated hydrocarbon solvents, will require consideration of applicable OSHA, EPA, and other federal, state, and local regulations.

No. 12

(Cream)

Rocsol Wax COC	12.00
Crodolene	1.00
Morpholine	0.80
Water (deionized)	45.00
White Spirit	20.00
Petroleum Spirit SBP 5	12.00
Silicone Fluid (300 cs)	5.00
Silicone Fluid (12,500 cs)	2.00
Snow Floss	4.00
Formalin	0.20

Procedure:

Melt wax at 100 C, add **Crodolene** and silicone and run in solvent. Heat water to 90 C, add morpholine and run into oil phase with efficient agitation. Homogenize, add formalin and stir to cool. A corrosion inhibitor will be desirable if a tin plate container is used.

No. 13

A	**Veegum T**	1.0
	Water	47.0
B	**Hoechst Wax S**	1.3
	Hoechst Wax E	0.2
	Silicone #530 Fluid	0.5
	Silicone #531 Fluid	3.0
	Sorbitan Monostearate	1.0
	Mineral Spirits	36.0
C	**Snow Floss**	10.0
D	Preservative	q.s.

Procedure:

Add the **Veegum** to the water slowly, agitating continually until smooth. Heat to 80 C. Heat B to 75 C with stirring. Avoid open flame. Heat B to A with continuous agitation. Add C to A and B and continue mixing until cool. Add D.

Directions for use:
Apply a thin even coat with a damp cloth. Allow to dry. Wipe to a brilliant luster with a clean dry cloth.

	No. 14	No. 15
		(Aerosol)
Syncrowax ERL	1.50	2.00
Silicone Fluid (300 cs)	1.50	4.00
White Spirit	16.00	18.00
Crill 4	0.50	0.60
Crillet 4	0.40	1.40
Water (deionized)	60.80	74.00
Corrosion Inhibitor	q.s.	q.s.

Procedure:
Melt waxes, add silicone and solvent. Adjust to 80 C, add **Crill 4**.
Heat water to 80 C, add **Crillet 4** and corrosion inhibitor. Add water to oil phase, homogenize to cool. Propellant: Arcton or Butane.

No. 16

(Reburnishing)

Volpo 010	2.00
Kerosene	85.00
Cab-O-Sil M5	3.00
Neuberg Chalk	25.00

Procedure:
Dissolve **Volpo 010** in kerosene, add **Cab-O-Sil M5** and stir until dispersed. A soft gel will form. Then add the Neuberg chalk, continue stirring until dispersed.

No. 17

(Reburnishing)

Crodolene	4.00
Kerosene	50.30
Neuberg Chalk	38.00

Bentonite 38	4.00
Water (deionized)	6.00
Ammonia 0.88	0.70

Procedure:

All components except ammonia and water are weighed and stirred together, ammonia is added followed by water, and stirring continued until a smooth dispersion results.

Note:

To produce creams of stiffer consistency **Syncrowax AW1** may be employed as a partial replacement for the **Crodolene.**

No. 18

(Cationic)

Silicone (DC-200)	2.0
Arquad® 2C-75	0.25
Ethomeen® S/12	0.25
Kerosene	8.0
Water	89.5

Procedure:

The silicone, kerosene and emulsifiers are thoroughly mixed together. Water is added with constant agitation. The resulting emulsion is run through a homogenizer to yield a stable white opaque emulsion with a light viscosity.

Automobile Polish and Cleaner

Formula No. 1

(Aerosol)

A	Veegum	1.5
	Sodium Carboxymethylcellulose (med. visc.)	0.2
	Water	48.0
B	Carnauba Wax	2.0
	Beeswax	2.0

	Oleic Acid	3.0
	Silicone (350 cs)	5.0
C	Morpholine	2.0
	Water	14.3
D	Mineral Spirits	20.0
E	Abrasive	2.0
F	Preservative	q.s.

Procedure:

Dry blend the **Veegum** and CMC and add slowly to the water, agitating continually until smooth. Heat to 90 C. Melt B, stirring until uniform. Heat Part C to 60–70 C and add to B with mixing, holding temperature at 85–90 C. Heat D to about 70 C and add to B and C with mixing. Add A with rapid agitation. Continue mixing until cool. Add E to other components and mix until uniform. Then add F.

Aerosol package:

Concentrate 90%; Propellant 12/114 , 60/40 10%.

No. 2

(Paste)

A	**Veegum**	1.0
	Water	29.0
B	Carnauba Wax	5.0
	Beeswax	3.0
	Ceresin	3.0
	A-C Polyethylene 629	2.5
	Silicone (350 cs)	5.0
	Stearic Acid	6.0
C	Morpholine	2.0
D	Naphthol Spirits E-1	36.0
E	Abrasive	7.5
	Preservative	q.s.

Procedure:

Add the **Veegum** to the water slowly, agitating continually until smooth. Heat to 95 C. Melt B. Add C and maintain temperature at 90 C. Add D to B and C. Add A with vigorous agitation. When temperature reaches 50 C, add E and continue mixing until cool.

Auto Paste Wax/Cleaner

(Nonionic)

A	**Viscasil**® (10,000)	5.0
	Carnauba Wax (No. 2)	4.0
	Wax LP	1.8
	Wax E	7.2
	Ozokerite 77Y	2.0
B	Mineral Spirits	40.5
	Kerosene	27.5
C	**Kaopolite SFO**	12.0

Procedure:

Heat components of Part A to 90 C (194 F) and hold at this temperature until all waxes are melted. Preheat Part B to 60 C (140 F) to prevent precipitation of the waxes. Add Part B slowly to Part A with constant low speed agitation. Continue agitation after Part B addition is complete. Add Part C slowly to A–B mixture. Continue agitation until the temperature reaches 40–42 C (104–108 F). Pour into chilled containers. Circulating cool air across the container and paste surface will aid solidification.

Application:

Apply a thin even coat with a damp cloth using circular motion to achieve maximum cleaning. Allow the film to dry to a white haze and buff with a clean cloth. Turn the cloth frequently to obtain maximum ease of buffing.

Warning:

When solvents are used as described above, proper safety precautions must be observed. All solvents must be considered toxic and should be used only in well ventilated areas. Prolonged exposure to solvent vapors must be avoided. If flammable solvents are used, storage, mixing, and use must be in areas away from open flames or other sources of ignition. The selection of any solvent, particularly chlorinated hydrocarbon solvents, will require consideration of applicable OSHA, EPA, and other federal, state, and local regulations.

Automobile Spray Wax

Formula No. 1

(Nonionic, Cold)

Gulf Mineral Seal Oil	20
Emulsifier Four	15.25
Triton X-100 or equivalent	0.75
Butyl Cellosolve	3.0
Water	61

Note:
To produce higher performance or lower cost products, the percent actives of this formula can be increased or decreased.

No. 2

(Cationic)

Accoquat 2C75	20
Butyl Cellosolve	5
Mineral Seal Oil	25
Water	50

Directions for use:
This spray formulation can be readily diluted 4 or 5 to 1 with water by the car wash operator before charging to his spray unit.

Note:
A recommended mineral seal oil for blending in the above formulation is **Mentor 28**, produced by Exxon (Humble Oil and Refining Company).

No. 3

(Amphoteric, Buffable)

Tomah Emulsifier 4	10.0
Mineral Seal Oil	13.0
Tomah Amphoteric L	3.0
Water	44.0
Tomah Emulsion C-340	20.0
SWS-235 Silicone Emulsion	10.0

Procedure:
Charge the **Emulsifier 4**, mineral seal oil, and **Amphoteric L**. Mix well. Slowly add the water and mix for approximately 30 min. When homogeneous, add the Carnauba Wax Emulsion **C-340**, mix, and then add the **SWS-235** Silicone Emulsion. Dilute this final product 1:5 with water and inject 16 oz per car through a DEMA system.

No. 4

(Amphoteric, Foaming)

Tomah Emulsifier Four	10.0
Mineral Seal Oil	13.0
Tomah Amphoteric L	3.0
Water	44.0
Tomah Amphoteric L	30.0

Procedure:
Charge the **Emulsifier 4**, mineral seal oil and 3.0 percent **Amphoteric L**. Mix well. Slowly add the water while blending and mix for 30 min. Then slowly add the remaining (30 percent) **Amphoteric L** with blending.

This formulation is stable, beads well, will not cause problems with detailing and reclaim water, and can be used in existing spraywax arches. The carwash operation should dilute 5 gal of the above formulation to 55 gal with water, then inject 10 oz of the dilute foam wax onto each car through a 204A DEMA system.

No. 5

(Hot Carnauba)

This is produced by slowly blending 40 parts of 70-90 F water into 45 parts of the above auto spraywax. When the water has been added, mix in 15 parts of Tomah's **Emulsion C-340.** Package in lined container. The Car Wash will usually dilute 10 gal of hot carnauba wax to 55 gal with water, then inject 6-10 oz of the dilute wax onto each car.

Auto Vinyl Top Polish

(Nonionic-Cationic)

GE SM-2033	23.0
GE SM-2035	25.0
Armac T	0.4
Water	51.6
Formalin Preservative	as required

Procedure:
Dissolve **Armac T** in warm water, then blend in other ingredients with mild agitation.

Application:
Clean vinyl surface. Apply polish evenly with a clean, soft cloth and buff to a high gloss.

Furniture Polish

Formula No. 1

(Anionic, Cream for Pump Spray)

Syncrowax ERL	0.50
Crill 35	0.50
Crillet 35	2.00
White Spirit	2.00
Deodorized Kerosene	2.00
Water (deionized)	89.00

Silicone Fluid (350 cs)	4.00
Formalin	0.20
Perfume	0.20

Procedure:
Melt wax. Add emulsifiers, silicone oil and solvent. Adjust temperature to 75 C. Heat water to 75 C and run into oil phase with vigorous stirring. Quickly add formalin at 40 C. Homogenize and fill off at 30-35 C.

No. 2

(Anionic, Silicone Paste)

Syncrowax RLS	20.00
Syncrowax HRS	5.00
Candelilla Wax	5.00
Polyethylene AC8	10.00
Rocsol Wax COA	5.00
Paraffin Wax 135/145	55.00
White Spirit	375.00
Silicone Fluid (1000 cs)	5.00
Oil Orange E	0.05

Procedure:
Melt polyethylene and paraffin waxes. Add Candelilla wax, **Rocsol** wax and **Syncrowaxes**, adjust to 110 C (do not overheat). Add solvent and reheat to 70 C; homogenize (Silverson) to 60 C. Add perfume. Stir with normal paddle or gate stirrer to 40 C. Fill off. Refrigerate filled containers.

No. 3

(Anionic, Cream)

Syncrowax HGL	4.00
Paraffin Wax 135/145	2.00
Paraffin Wax 120/125	2.00
Rocsol Micro 10	2.00
Cithrol GMO N/E	2.80

White Spirit	37.50
Water (deionized)	100.00
Teepol 610	3.00
Formalin	0.50
Perfume	0.20

Procedure:

Melt waxes, add solvent. Adjust temperature to 75 C and add **GMO**. Heat water to 75 C, add **Teepol**. Run into oil phase with vigorous stirring (Silverson). Cool to 20 C with moderate stirring. Add perfume and formalin. Low temperature milling will improve the cream if an homogenizer has not been used during manufacture.

No. 4

(Anionic, Lavender Paste)

Syncrowax RLS	15.00
Candelilla Wax	5.00
Polyethylene AC8	10.00
Rocsol Wax COA	8.00
Rocsol Wax B	2.00
Paraffin Wax 135/145	60.00
Oil Violet	0.02
White Spirit	350.00
Coumarin	0.50
Lavender Perfume	5.00

Procedure:

Melt waxes, do not overheat (115 C). Start stirrer/homogenizer. Run cold solvent into molten wax, add dye and perfume. Stop homogenizer when polish has reached 50 C but continue with moderate stirring. Fill into containers between 40-45 C. Refrigerate or fill in a cool room. For large containers fill by two or three separate additions.

No. 5

(Anionic, Cream with Silicone)

Syncrowax HGL	3.00
Carnauba Wax	1.00
Silicone Fluid (350 cs)	1.50
Crill 4	2.30
Crillet 4	0.50
Perfume	0.20
White Spirit	20.00
Water (deionized)	71.50
Formalin	0.20

Procedure:

Melt waxes, add silicone fluid and adjust to 75 C. Combine **Crill, Crillet,** and water; heat to 75 C. Add water, heat to 75 C. Add water to oil phase with vigorous stirring (Silverson). Cool to 40 C. Add perfume and formalin.

No. 6

(Nonionic, W/O Aerosol)

		% in Aerosol
Concentrate:		80.0
A VM & P Naphtha	14.5	
Span 90	0.7	
B **Co-Wax*** (10% emulsion)	16.2	
Dow Corning 922 Emulsion	8.0	
Water	60.1	
C Perfume	0.5	
Propellant:		20.0
Isotron 11	50.0	
Isotron 12	50.0	

* Co-Wax is supplied in 100% form. It is dispersible to a 10% emulsion by stirring in 180–190 F (82–88 C) water.

Procedure:
Mix Parts A and B separately. Then add B to A with high-shear mixing. Add perfume and pressure-fill.

Directions for use:
Shake well. Hold can 12-14 in. from surface to be sprayed.

Package:
In lacquer-lined cans with appropriate valve and button.

Warning:
Contents under pressure. Do not puncture. Exposure to heat or prolonged exposure to sun may cause bursting. Do not throw into fire or incinerator. Keep from children.

Note:
Good gloss and low smear polish lightly scented with lemon odor.

No. 7

(Nonionic)

Clindrol 200-S	3
Stearic Acid	1.2
Mineral Seal Oil	20
Heavy Mineral Oil	20
Water	55.8

Procedure:
Clindrol 200-S and stearic acid are dissolved in the oil by heating to 80 C. This solution is slowly added with rapid stirring to water previously heated to the same temperature (70-80 C). The final product exhibits a consistency well suited to a furniture polish in that it pours readily, but at the same time permits control in confining the fluid to the desired area of application.

No. 8

(Nonionic, Silicone)

Rocsol Wax CGO	2.00
Stearyl Alcohol	1.00
Volpo CO20	2.00

Silicone Oil	2.00
White Mineral Oil	4.00
Water (deionized)	84.00

Procedure:

Melt together the wax, stearyl alcohol and emulsifier. Add the mineral spirits containing the silicone, and finally the water. Rapid stirring should be used in the latter operation and this should be continued until the batch is cooled. A thin cream results, having a good shelf life and imparting an instant high gloss to furniture. Thicker creams can be prepared by increasing the quantity of stearyl alcohol. If a wipe gloss polish is required, the mineral spirits may be replaced by a more volatile petroleum ether type of solvent.

No. 9

(Nonionic, Waxless)

Volpo O5	0.15
American Turpentine	25.00
White Mineral Oil	25.00
Ethyl Alcohol (74 OP)	25.00
Crodolene	0.50
Water (deionized)	50.00
Morpholine	0.18

Procedure:

Blend alcohol and water, morpholine, **Crodolene** and **Volpo**. Add oil and turpentine cold and homogenize.

No. 10

(General Purpose, Liquid)

Carnauba Wax	0.50
Silicone Oil (500 cs)	4.00
Pine Oil, Dipentene, or Terpinolene	1.00
White Spirit	4.00
Crillet 35	2.20
Crill 35	0.60
Water (deionized)	87.70

Procedure:

Heat all components except water to 85 C. Heat water to 85 C. Add water to oil under high speed mixer until temperature falls to 40 C and then homogenize. Fill off at 30 C.

No. 11

(Silicone, Liquid, Waxless)

Silicone Oil (350 cs)	4.50
White Spirit	4.50
Crillet 35	2.00
Crill 35	0.50
Water (deionized)	88.50

Procedure:

Heat all components except water to 85 C. Heat water to 90 C. Add water to oil under high speed mixer, stir until cool (30-40 C), homogenize.

No. 12

(Spray, Silicone)

		% in Aerosol
Concentrate:		50.0
J-324 Wax Dura	1.2	
Paraffin Wax (low melting 50-52 C)	0.6	
Cardis Polymer 10	1.0	
Silicone Oil 200 (visc. 350)	6.0	
Stoddard Solvent	91.2	
Propellant:		50.0
Isotron 12	50.0	
Isotron 11	50.0	

Procedure:

Warm ingredients with stirring in a steam-jacketed container until waxes are melted and a homogeneous liquid is obtained. Cool rapidly with stirring and pressure fill.

Directions for use:

Shake well. Spray surface lightly from a distance of 12-14 in. Use a clean, dry cloth to spread wax evenly. After several minutes, buff to desired luster.

Package:

Tinplate container with paint-type valve.

Warning:

Contents under pressure. Do not puncture. Exposure to heat or prolonged exposure to the sun may cause bursting. Do not throw into fire or incinerator. Keep from children.

Solventless Liquid Polish

(Anionic, For Plastic Laminates and Finishes)

Syncrowax HGL	4.00
Crosterene	2.70
Silicone Fluid (500 cs)	3.00
Water (deionized)	89.00
Triethanolamine	1.30
Preservative	q.s.

Procedure:

Melt waxes and silicone. Cool to 75 C. Heat water to 75 C, add triethanolamine to waxes and then run in water with vigorous stirring.

Furniture and Floor Polish

Formula No. 1

(Anionic)

Syncrowax HGL	6.00
Syncrowax HRS	10.00
Polyethylene AC 8	10.00
Rocsol COA	8.00
Rocsol B	2.00
Paraffin Wax 135/145	64.00

Oil Orange E	0.10
White Spirit	350.00
Perfume	2.00

Procedure:

Melt waxes, do not overheat (115 C). Start stirrer/homogenizer. Run cold solvent into molten wax, add dye and perfume. Stop homogenizer when polish has reached 50 C but continue with moderate stirring. Fill into containers between 40–45 C. Refrigerate or fill in a cool room. For large containers fill by two or three separate additions.

Note:

Paraffin wax 135/145 can be either 135/140 or 140/145.

In practice, however, desirable properties can be obtained by mixing both paraffin waxes in equal proportions.

No. 2

(Anionic, Liquid, Solvent-Based)

Rocsol Micro 8	3.00
Syncrowax BB2	3.00
Paraffin Wax 140/145	6.00
Syncrowax HRS	3.00
White Spirit	150.00

Procedure:

Heat waxes to 110 C. Add solvent. Silverson down to 40 C.

No. 3

Rocsol Wax COA	4.00
Beeswax	1.00
Ceresine 140/145	2.50
Paraffin Wax 135/140	22.40
Dyestuff	0.10
White Mineral Spirits	100.00

Procedure:
Melt the waxes, add the dyestuff and then the solvent. Pour into tins at 45 C. Silicone oil may then be incorporated, if desired.

Floor Polish

Formula No. 1

(Anionic)

Primal B 336 (15%)	72.50
Schenectady SR88 (15%)	7.50
Syncrowax Emulsion (15%)	20.00
Carbitol	1.50
Ethylene Glycol	1.50
Tributoxy Ethyl Phosphate	0.35
FC 128 (1% sol'n.)	6.40
Dowicil 100	0.20

Procedure:
Prepare **Schenectady SR88** resin solution and **Syncrowax** emulsion as for previous formulations. Either wax A or B can be used in this product.

Blend polymer, resin solution containing tributoxy ethyl phosphate and the selected wax emulsion, then add Carbitol, ethylene glycol, **FC 128** solution and **Dowicil 100**. Stir gently for 30 min to ensure an homogeneous blend.

No. 2

(Anionic)

Syncrowax RLS	17.00
Syncrowax HGL	3.00
Syncrowax HRS	10.00
Rocsol Wax COA	5.00
Paraffin Wax 125/130	65.00
Oil Orange E	0.05
White Spirit	270.00
Perfume	0.50

Procedure:
   Melt waxes at 110 C, add solvent and dyes; stir to finish at 70 C. Cool to 50 C. Add perfume and fill at 40–45 C. This product is improved by the use of an homogenizer while cooling from 70 C to 50 C but not below. A Silverson homogenizer is recommended.

	No. 3	No. 4
	(Anionic, Nonbuffable)	*(Anionic, Semibuffable)*
RWL 100 Latex	35.00	20.00
Schenectady Resin SR88 (15% ammoniacal sol'n.)	20.00	20.00
Syncrowax Emulsion (15%)	20.00	35.00
Tributoxy Ethyl Phosphate	0.70	0.40
FC 128 (1% sol'n.)	1.20	1.20
Dowicil 100	0.20	0.20
Water (deionized)	22.90	23.20

Procedure:
   Method of manufacture for Resin Solution:

Schenectady SR88	15.00
Ammonia SG 0.88	3.80
Water (deionized)	81.20

   Heat the water to 80 C and add half the ammonia. While stirring slowly add the **SR88**. Maintain temperature and add remainder of ammonia, stir until solution is complete.
   The tributoxy ethyl phosphate is added to the appropriate final formulation quantity of cool resin solution and stirred; additional ammonia may be required to complete this solution. The **SR** resin solution acts as a levelling agent during both application and drying. The tributoxy ethyl phosphate is a plasticizing agent for the resin.

   Method of manufacture for Wax Emulsion:

Syncrowax BE1	13.00	—
Syncrowax BE14	—	13.00
Crodolene	1.80	0.60

2-Amino-2-Methyl Propanol (AMP)	–	0.60
Morpholine	2.20	–
Emulsifier ET 0650	–	2.50
Water (deionized)	83.00	83.30

Melt wax, add **Crodolene** and **Emulsifier ET 0650** and adjust temperature to 98 C. Add the amine and allow to react for 3 min. While stirring the molten wax mix, add half the water (preheated to 90 C) in a steady stream. The melt will gel during the initial water addition; careful stirring will maintain an homogeneous mix. When all the hot water has been added, the remainder of the water is added in a steady stream. Gently stir until cool.

Metal Polish

Formula No. 1

(Cationic)

A	Kerosene	1.0
	Perchloroethylene	15.0
	Arquad® 2C-75	2.0
	Ethomeen® S/12	1.0
	Silicone Oil (200–500 cp)	4.0
B	Water	53.0
	Silica Flour	10.0

Procedure:

Mix and heat Part A and Part B separately until uniform. Add A and B together and cool.

No. 2

Condensate PHL	4.0
Silica	17.8
Water	78.0
Cottonseed Fatty Acid	0.2

Silver Polish

Formula No. 1

(Nonionic, Liquid)

A	Veegum	2.0
	Kelzan	0.3
	Water	76.7
B	Snow Floss	15.0
C	Triton X-102	5.0
D	Nacap	1.0
E	Perfume and Preservative	q.s.

Procedure:

Dry blend the **Veegum** and the **Kelzan**. Add to the water slowly, agitating continually until smooth. Add B to A. Mix well. Add C to A and B. Add D to other components and mix until well dispersed. Add E.

Directions for use:

Pour a small amount of cleaner onto a damp cloth and clean article with moderate rubbing. Rinse with water, dry and polish with a separate cloth.

No. 2

(Nonionic, Liquid)

A	Veegum	2.0
	Water	68.9
B	Triton X-102	5.0
	Silicone (350 cs)	2.0
	Ammonia	2.0
	Abrasive	20.0
	Rotax	0.1
C	Preservative	q.s.

Procedure:

Add the **Veegum** to the water slowly, agitating continually until smooth. Add the components in B in the order listed with mixing after each addition.

No. 3

(Nonionic, Liquid)

A	Veegum	2.0
	Sodium Carboxymethylcellulose (low visc.)	0.1
	Water	76.9
B	Snow Floss	15.0
C	Buffer Sol'n.[1]	q.s.
D	Triton X-102	5.0
E	Octadecyl Thioglycolate	1.0
	Perfume and Preservative	q.s.

---

[1] Buffer solution: 60 parts–1M citric acid.
40 parts–saturated sodium citrate sol'n.
(87 g/100 cc) filtered

Procedure:
Dry blend the **Veegum** and the CMC. Add to the water slowly, continually agitating until smooth. Add B to A. Buffer to pH 5.0 and add D. Add E and mix until well dispersed.

Directions for use:
Pour small amount of cleaner onto a damp cloth and clean article with moderate rubbing. Rinse with water, dry and polish with a clean dry cloth.

No. 4

(Amphoteric)

Miranol® DM	50.0
Kaopolite SF-L or **73**	50.0

Procedure:

Charge 40 parts of **Miranol® DM** to a suitable paste mixer. Run the mixer continuously while adding **Kaopolite**. Add the remaining 10 parts of **Miranol® DM** in small increments to avoid overloading the mixer drive. The object is to blend at the heaviest practicable consistency to provide thorough dispersion. Continue mixing until fully dispersed and smooth.

This formulation is very effective on most ornamental metals in the home such as brass, copper, pewter, nickel, stainless steel, aluminum, etc., when small amounts of ammonia are added to it.

Shoe Polish

Formula No. 1

(Anionic, Cream)

Syncrowax HGL	6.00
Syncrowax AW1	2.60
Paraffin Wax 140/145	8.00
White Spirit	28.00
Stearyl Alcohol	1.20
Water (deionized)	52.20
Sodium Silicate	0.30
Morpholine	1.20
Color (water soluble)	0.50

Procedure:

Melt waxes and stearyl alcohol to 90-100 C. Add white spirit and adjust temperature to 77 C. Heat water to 77 C, add morpholine, sodium silicate and color. Add wax to water phase under an ordinary mixer, stir to 45-50 C. Add a suitable polish perfume.

No. 2

(Anionic, Waterproof)

Syncrowax RLS	10.00
Syncrowax HRS	15.00
Rocsol Micro 10	5.00

Syncrowax BB2	2.50
Polyethylene Wax AC6	5.00
Paraffin Wax 135/140	60.00
Lanolin Crodapur	2.50
White Spirit	2.75
Oil Orange E	0.30

Procedure:

Melt waxes and lanolin. Run in hot solvent at 90 C under vigorous stirring. Add dyestuffs and perfume when cool.

No. 3

(Anionic, Black Paste)

Syncrowax HGL	20.00
Polyethylene AC6	5.00
Rocsol Wax CFD	2.00
Paraffin Wax 135/145	73.00
Nigrosine Base NB	5.00
Syncrowax AW1	5.00
Lime	1.00
Water (deionized)	10.00
Teepol 610	0.10
White Spirit	300.00

Procedure:

Melt waxes and adjust temperature to 100 C. Add **Syncrowax AW1** followed by Nigrosine base, stir for 15 min at 100 C to ensure solution of dyestuff. Add lime suspender in a little white spirit. Continue the reaction at 115 C for 15 min to form the calcium soap of the wax acid. Add portion of white spirit. When temperature reaches 100 C, add water/**Teepol** mixture and rest of solvent. Stir until cool. Fill at temperature appropriate to requirements. Lightly refrigerate.

No. 4

(Anionic, Black Paste)

Syncrowax HGL	7.00
Polyethylene AC6	3.00
Rocsol Wax COB	18.50

Rocsol Wax B	5.00
Syncrowax BB1	2.50
Paraffin Wax 135/145	64.00
Nigrosine Base NB	5.00
Syncrowax AW1	4.00
White Spirit	260.00

Procedure:

Melt waxes at 95-100 C. Add **Syncrowax AW1** and Nigrosine, stir for 15 min to ensure solution of dyestuff. Add shite spirit and stir to cool. A Silverson type stirrer will improve the texture of this polish. Fill at temperature appropriate to requirements.

Note:

Nigrosine base predissolved in equal weight of oleine can be used. This dye mixture should be added to the polish when the temperature has reached 45 C after the addition of all the solvent.

No. 5

(Anionic, Brown Paste)

Syncrowax HGL	30.00
Syncrowax HRS	10.00
Paraffin Wax 135/145	60.00
Asphaltum	5.00
Oil Brown W	3.00
White Spirit	250.00
Perfume	3.00

Procedure:

Melt waxes (100-110 C). Add asphaltum and stir for 5 min to disperse. Run in solvent, add dyestuff. Silverson to 45 C. Fill at temperature appropriate to requirements.

No. 6

(Anionic, Brown Paste)

Syncrowax RLS	25.00
Polyethylene AC6	5.00
Rocsol Wax COB	5.00

Paraffin Wax 135/145	65.00
Oil Brown W	3.00
White Spirit	278.00

Procedure:

Melt waxes (100–110 C). Run in solvent and add dyestuffs. Stir (Silverson) to cool to 45 C. Fill at temperature required to give appropriate paste properties.

	No. 7	No. 8
	(Anionic, Self-Shine)	
Syncrowax BE14	13.00	—
Syncrowax HGL	—	6.00
Rocsol Wax CFD	—	6.00
Polyethylene Wax AC 629		4.00
Crodolene	0.60	0.90
AMP	0.60	0.90
Emulsifier ET 0650	2.50	2.50
Water (deionized)	82.10	78.50
Color	0.70	0.70
Perfume	0.50	0.50

Procedure:

Melt wax, add **Crodolene** and adjust to 98 C. Add AMP and react for 3 min. Heat half of the water to 90 C and add in a steady stream to the above mixture under a stirrer. Reheat if necessary to entirely melt the wax. Add remainder of water steadily and then stir until cool.

No. 9

(Nonionic, White)

A	**Veegum**	1.0
	Sodium Carboxymethylcellulose (med. visc.)	0.5
	Water	77.7
B	**Darvan No. 1**	0.3
C	Dibutyl Phthalate	3.0
	Titanium Dioxide	15.0

Neatsfoot Oil	2.0
Triton X-102	0.5
D Preservative	q.s.

Procedure:

Dry blend the **Veegum** and CMC and add to 30 parts of the water slowly, agitating continually until smooth. Mix B in the remaining water and add to A. Add the components in C in the order listed to A and B and mix until uniform. Then add D.

No. 10

(White)

Condensate PS	3.0
Titanium Dioxide	36.0
Water	61.0
Gum Arabic	1.0-1.5

Shoe Cleaner

(Nonionic, For Suede and White)

		% in Aerosol
Concentrate:		85.0
Water	67.5	
Butyl Cellosolve	2.0	
Triton X-100	20.0	
Deodorized Kerosene	9.0	
Lanogene	1.3	
Perfume	0.2	
Propellant:		15.0
Isotron 12	40.0	
Isotron 114	60.0	

Procedure:

Use the water to dissolve the other components of the concentrate. Mix well. Product should be pressure filled.

Directions for use:

Foam a small amount on surface to be cleaned. Use a soft cloth to spread evenly. A brush may be used on stubborn stains and scuff marks. A wire brush should be used to restore the nap on suede leather after drying.

Package:

Lacquer-lined container with a foam-type valve.

Warning:

Contents under pressure. Do not puncture. Exposure to heat or prolonged exposure to sun may cause bursting. Do not throw into fire or incinerator. Keep from children.

Chapter IV

GENERAL INDUSTRIAL CLEANERS

High Alkaline Cleaner

Formula No. 1

(Anionic)

Ninex 24	5-7
Sodium Metasilicate Pentahydrate	10
Potassium Hydroxide	3
Water	balance

Note:

May be used as a wax stripper, steam cleaner, whitewall tire cleaner. Sometimes a cloud will form upon standing overnight. This is due to colloidal silica and can be settled out or filtered off.

No. 2

(Anionic, Heavy Duty)

Alkaphos-10	10
Tetrapotassium Pyrophosphate	10
Metso (anhydrous)	5
Potassium Hydroxide	5
Water	70

Note:

In use this product is diluted with about 4 or 5 volumes of kerosene, then sprayed or rubbed on to the greasy parts and the loosened soil finally rinsed away with water.

	No. 3	No. 4
		(Anionic)
Alkaphos-6	6	4
Sodium Tripolyphosphate	5	–
Metso (anhydrous)	–	10
Potassium Hydroxide (45%)	2	–
Water	83	86

Note:

Both these cleaners are liquids and can be diluted 4 oz/gal of water for floor and wall cleaning. Foam is moderate, and the formulated products are nonrusting and can be contained in plain steel drums.

No. 5

(Anionic, Degreaser)

Aromatic 150	40.0
Monamulse 947	20.0
NaOH (50%)	4.0
Water	36.0

Procedure:

Add the ingredients in the order listed with good agitation. After the product has cleared, continue the agitation for at least 15 min to assure maximum product stability and uniformity.

Use concentration:

Truck and rail cars: ½-1 qt/gal of water
Heavy duty cleaning: approximately 2 qt/gal of water
Rinse with either cold or hot water or steam.

Note:

This formulation has been stored at 40 F and room temperature for several months and remains an essentially clear, opalescent liquid.

No. 6

(Anionic)

Water	77.50
STPP	5.00

TSP	5.00
Metso	5.00
ESI-Terge 320	7.50

Procedure:

Add in order listed with adequate agitation allowing powders to dissolve completely before adding **ESI-Terge 320**.

Specifications:

pH:	12.5–13.5
Active:	22.5%
Solids:	22.5%
Visc.:	25 cps max LV #1 spindle @ 60 rpm @ 25 C

No. 7

(Nonionic-Anionic, Dry-Type)

Plurafac RA-20 Surfactant	2
Klearfac AA-270 Surfactant	1
Na_2CO_3	47
Na_3PO_4	30
$Na_5P_3O_{10}$	20

No. 8

(Nonionic-Anionic, Heavy-Duty Powder)

Sodium Metasilicate (anhydrous)	31.7
Sodium Hydroxide Flakes	31.7
Sodium Carbonate	31.6
Monaterge 85	5.0

This formulation remains free flowing and discolors only slightly after storage at 20 C and 50 C.

No. 9

(Nonionic-Anionic, Heavy-Duty Liquid)

Water	54.7
Tetrapotassium Pyrophosphate (60%)	25.3

Sodium Hydroxide (50%)	10.0
Monaterge 85	10.0

Procedure:
Add ingredients in the order listed with good agitation.
pH (as is) = 13.7, and foam is moderate.

	No. 10	No. 11
	(Nonionic-Anionic,	
	Heavy Duty Liquid)	
Water	45.0	40.0
N Sodium Silicate	40.0	40.0
Sodium Hydroxide (50%)	10.0	10.0
Monaterge 85	5.0	10.0
Surface tension: (dynes/cm—		
2% sol'n. @ 20 C)	30.5	30.3
pH (as is) = 12.7		
Foam: Moderate		

Procedure:
Charge water, start agitation and then add N sodium silicate. Then add NaOH solution and agitate until clear. Finally add **Monaterge 85** and continue agitation (approx. 45 min) until solution clears.

Use Dilutions:
Medium Duty: 2-4 oz/gal
Heavy Duty: 4-8 oz/gal

Note:
Formula 10 is a good starting formulation for general all purpose cleaning. Formula 11 is recommended where heavier soil loads are encountered.
Formula 10 is an upper limit formulation. Formulas containing lower concentrations of NaOH and/or N sodium silicate will usually clear faster. A very light, fine precipitate will be present in the final product, as is typical of silicate containing systems. If hazy N sodium silicate is used the finished formula will be hazy. This haze will eventually settle out, resulting in a clear product with a fine white precipitate.

No. 12

(Nonionic-Anionic)

		Mixing Order
Ninol 1301	5	5
Sodium Metasilicate Pentahydrate	10	2
Tetrapotassium Pyrophosphate	5	3
Stepanate X	10	4
Water	70	1

Procedure:
Dissolve phosphate and silicate in water. Add the **Stepanate X** and **Ninol 1301**. Stir until uniform.

pH:	12.5
Gardner Color:	1
Brookfield Visc.:	About 5 cps @ 25 C
Appearance:	Clear, slight yellow liquid.

No. 13

(Nonionic-Anionic)

Ninol 1301	10
Sodium Metasilicate Pentahydrate	10
Tetrapotassium Pyrophosphate	5
Stepanate X	8
Water	67

Note:
More alkali can be incorprated if the proportion of **Stepanate X** (xylene sulfonate coupler) is increased. This formulation is not viscous, but if a higher consistency is desired then a different coupler, **Stepanol B-153**, is recommended.

No. 14

(Amphoteric, Liquid)

Miranol® J2M Conc. or **Miranol® J2M-SF Conc.**	2.0
Water	38.0

| Potassium Hydroxide (45%) | 10.0 |
| Kasil #1 | 50.0 |

No. 15

(Amphoteric, Powder)

Miranol® J2M Conc. or Miranol® J2M-SF Conc.	1.0- 2.0
Sodium Gluconate	6.0- 6.0
Sodium Hydroxide (flakes)	93.0-92.0

Note:
This formulation can be used for bottle washing.

No. 16

(Amphoteric, Dry Powder)

Caustic Soda	90
Soda Ash	5
Monateric LF-100	5

Monateric LF-100 has outstanding stability on caustic soda. This formula has been stored for weeks at 95 C and for several months at 50 C with little or no change in color or flowability. Detergency and foaming properties were also unaffected.

No. 17

(Amphoteric, Liquid)

Monateric CA (35%)	10.0
Sodium Metasilicate (5H$_2$O)	20.0
Sodium Hydroxide	3.3
Water	66.7

No. 18

(Amphoteric, Heavy Duty)

Alkateric 2CIB	8-10
Sodium Metasilicate Pentahydrate	20-25
Sodium Hydroxide	3- 5
Water	to 100

Acid Cleaner

Formula No. 1

(Anionic–Amphoteric, Heavy Duty)

Alkateric 2CIB	3- 5
Phosphoric Acid (85%)	50-70
Alkasurf LA-Acid	3- 5
Water	to 100

No. 2

(Nonionic-Amphoteric)

Water	78.0-50.0
Monaterge LF-945	2.0- 5.0
Sulfuric Acid	20.0-40.0
Hydrofluoric Acid	0- 5.0

Note:

Depending on soil load, use concentrations of **Monaterge LF-945** as low as 0.05% (with 1% sulfuric acid) produced completely clean, water-break-free surfaces.

No. 3

(Amphoteric, Liquid)

Monateric Cy Na (50%)	2.0
Gluconic Acid (50%)	6.0
Phosphoric Acid	54.7
Water	37.3

All-Purpose Cleaner

Formula No. 1

(Anionic, Heavy Duty)

Water	84.0
TSP	5.0

Sodium Metasilicate	5.0
ESI-Terge 320	4.0
D.D.B.S.A.	2.0

Procedure:

Dissolve TSP and sodium metasilicate in water. Add **ESI-Terge 320** and D.D.B.S.A. Agitate until clear.

Specifications:

pH:	12-13
Activity:	16%
Solids:	16%
Visc.:	120 cps LV #1 spindle 60 rpm @ 25 C

Note:

To convert to a wax stripper, 3-5% ammonia or monoethanolamine is added to this cleaner.

No. 2

(Anionic)

Sulframin® 85 Powder	1.0
Sodium Sesquicarbonate	46.1
Trisodium Phosphate	8.5
Tetrasodium Pyrophosphate	17.8
Sodium Sulfate	26.6

No. 3

(Anionic, Nonphosphate)

Water	85.0
Caustic Soda Flakes	0.7
Monamine ALX-100S	6.0
Butyl Cellosolve	5.3
Sodium Carbonate	2.0
Monaquest CA-100	1.0

Procedure:

Add ingredients in order listed.

No. 4

(Nonionic, Spray-Type)

Neodol 23-6.5	1.7
Cocodiethanolamide	0.5
Trisodium Phosphate	1.0
Sodium Metasilicate $5H_2O$	1.7
Butyl Oxitol®	3.5
Water, Dyes, Perfume, etc.	q.s.

No. 5

(Nonionic, Janitor-in-a-Drum Type)

Neodol 23-6.5	5.0
Neodol 25-3	2.5
Butyl Oxitol	6.0
Pine Oil	0.25
Tetrapotassium Pyrophosphate	3.0
Sodium Metasilicate $5H_2O$	2.0
Sodium Xylene Sulfonate (40% AM)	1.0
Water	q.s.

	No. 6	No. 7
	(Nonionic, Liquid, Disinfectant)	
Neodol 23-6.5[1]	1	2
Didecyldimethylammonium Chloride[2] (50% AM)	5	10
Neodol 25-9	5	7.5
Tetrapotassium Pyrophosphate	4	8
Trisodium Phosphate	2	–
Butyl Oxitol	5	–
Water	79	74.5
Appearance	clear	clear
1/64 Use Dilution	clear	clear

---

[1] Note:
  Cleaners making a disinfectant claim require EPA registration. If necessary, Shell

can provide additional information on **Neodol** products to assist users in obtaining registration.
[2] **Bardac 22**, 50%w in IPA (20%) and water (30%), Lonza, Inc.

No. 8

(Nonionic)

A **Veegum T**	0.45
Kelzan	0.15
Water	72.40
B **Monamid 150-ADD**	0.50
Tetrapotassium Pyrophosphate	2.00
Water	21.00
Plurafac C-17	2.50
C Ammonium Hydroxide (28%)	1.00

Procedure:
Dry blend the **Veegum T** and **Kelzan** and add to the water slowly, agitating continuously until smooth. Combine B, stirring slowly to dissolve the tetrapotassium pyrophosphate (avoid incorporation of air). Add B to A with mixing. Add C with stirring.

Note:
This cleaner is ready to use and should not be diluted. It may be packaged in a pump dispenser bottle.

No. 9

(Nonionic-Anionic, Liquid Concentrate)

Neodol 25-3S (60% AM)	5
Neodol 25-9	5
Neodol 25-3	2.5
Butyl Oxitol	6
Pine Oil	0.25
Tetrapotassium Pyrophosphate	3
Sodium Silicate Pentahydrate	2
Ultra SXS	1
Water	q.s.

Visc.: 5 cp Clearpoint, 3 C; pH 11.8

Note:
Appearance at 1/64 aqueous dilution is clear. Recommended use concentrations: As is for tough jobs. 1/4 dilution for heavy work and 1/64 dilution for most applications.

	No. 10	No. 11
	(Amphoteric, Liquid Concentrate)	
Water	34.3	49.8
Potassium Hydroxide (45%)	8.6	6.7
Kasil #1	25.7	19.2
Tetrapotassium Pyrophosphate	13.3	10.0
Miranol® J2M Conc.	6.7	5.7
Potassium Carbonate	11.4	8.6
Solids	40%	30%
Crystallizing temperature	3 F	12 F

Procedure:
Mix in the order listed. Some turbidity may be noted during the mixing of the potassium carbonate, which will disappear upon standing.

No. 12

(Amphoteric)

Miranol® JEM Conc.	1.0- 3.0
Sodium Metasilicate Pentahydrate	9.0- 9.0
Tetrapotassium Pyrophosphate	20.0-20.0
Water	70.0-68.0

pH: 13.4

No. 13

(Amphoteric)

Miranol® JEM Conc.	1.0- 3.0
Sodium Metasilicate Pentahydrate	8.5- 8.5

Tetrapotassium Pyrophosphate	15.0-15.0
Tetrasodium Pyrophosphate	10.0-10.0
Water	65.5-63.5

pH: 13.5

No. 14

(Amphoteric)

Miranol® JEM Conc.	1.0- 3.0
Sodium Metasilicate Pentahydrate	8.5- 8.5
Tetrapotassium Pyrophosphate	17.0-17.0
Tetrasodium Pyrophosphate	8.0- 8.0
Water	65.5-63.5

pH: 13.4

No. 15

(Amphoteric)

Miranol® JEM Conc.	3.0- 3.0
Starso	38.0-56.0
Potassium Hydroxide (45%)	10.0-25.0
Water	49.0-16.0

pH: 13.0

	No. 16	No. 17
	(Amphoteric)	
Antaron FC-34	25	15
Sodium Metasilicate	5	20
Cheelox® B-13	1	1
Water	69	64

	No. 18	No. 19
		(Amphoteric)
Igepal CO-710	10	5-10
Antaron FC-34	3	3
Tetrasodium Pyrophosphate and/or		
Trisodium Phosphate	—	1- 3
Sodium Nitrite	0.5	0.5
Water	86.5	90.5-83.5

No. 20

(Amphoteric, Janitorial)

Miranol® C2M-SF Conc.	10.0
Tetrapotassium Pyrophosphate	30.0
Star	16.0
Water	44.0

Note:

This formulation may be used as a concentrate, packed in 1/4 to 1/2 oz packages (light plastic in aluminum foil) for use in 10 to 12 quart pails of water. It may be used for floors, walls, painted surfaces and appliances.

No. 21

(Amphoteric)

Tetrapotassium Pyrophosphate (TKPP)	5.0
Tetrasodium Pyrophosphate (TSPP)	5.0
Water	66.0
Potassium Hydroxide (45%)	4.0
Tall Oil Fatty Acid	10.0
Miranol® C2M-SF Conc.	10.0

No. 22

(Amphoteric, Sanitizing, Light Duty)

Miranol® C2M-SF Conc.	15.0
Sodium Tripolyphosphate	3.0
Trisodium Phosphate	3.0

Quaternary Ammonium Salt Germicide (50% active)	2.0
Organic Sequestering Agent (30%)	1.0
Water	76.0

Note:

Where high water hardness is encountered, it is suggested that one per-cent of an organic sequestering agent be used for each 100 ppm water hardness.

No. 23

(Amphoteric)

Miranol® JEM Conc.	1.0
Sodium Metasilicate Pentahydrate	9.0
Tetrapotassium Pyrophosphate	20.0
Water	70.0

pH: 13.4

No. 24

(Amphoteric)

Miranol® JEM Conc.	3.0
Sodium Metasilicate Pentahydrate	9.0
Tetrapotassium Pyrophosphate	20.0
Water	68.0

pH: 13.4

No. 25

(Amphoteric)

Miranol® JEM Conc.	3.0
Sodium Metasilicate Pentahydrate	8.5
Tetrapotassium Pyrophosphate	15.0
Tetrasodium Pyrophosphate	10.0
Water	63.5

pH: 13.5

No. 26

(Amphoteric)

Miranol® JEM Conc.	3.0
Sodium Metasilicate Pentahydrate	8.5
Tetrapotassium Pyrophosphate	17.0
Tetrasodium Pyrophosphate	8.0
Water	63.5

pH: 13.4

No. 27

(Amphoteric)

Miranol® JEM Conc.	2.0
Sodium Metasilicate Pentahydrate	9.0
Tetrapotassium Pyrophosphate	20.0
Water	69.0

pH: 13.4

No. 28

(Amphoteric)

Miranol® JEM Conc.	1.0
Sodium Metasilicate Pentahydrate	8.5
Tetrapotassium Pyrophosphate	15.0
Tetrasodium Pyrophosphate	10.0
Water	65.5

pH: 13,5

No. 29

(Amphoteric)

Miranol® JEM Conc.	1.0
Sodium Metasilicate Pentahydrate	8.5
Tetrapotassium Pyrophosphate	17.0

| Tetrasodium Pyrophosphate | 8.0 |
| Water | 65.5 |

pH: 13.4

No. 30

(Amphoteric)

Miranol® JEM Conc.	3.0
Starso	56.0
Potassium Hydroxide (45% liquid)	10.0
Water	31.0

pH: 13.2

No. 31

(Amphoteric, Concentrate)

Miranol® C2M-SF Conc.	10.0
Tall Oil Fatty Acid	6.0
Potassium Hydroxide (45%)	1.8
Ethylenediamine Tetra-Acetic Acid	2.0
Pine Oil	1.0
Ammonia (28%)	5.0
Sodium Tripolyphosphate	5.0
Sodium Dichromate	0.1
Water	69.1

No. 32

(Amphoteric, Concentrate)

Miranol® C2m-SF Conc.	10.0
Butyl Cellosolve	5.0
Sodium Carbonate	4.0
Ethylenediamine Tetra-Acetic Acid	2.0
Sodium Xylene Sulfonate	4.0
Water	75.0

No. 33

(Amphoteric-Anionic)

Miranol® C2M-SF Conc.	10.0
Foamole AR	6.0
Ultrawet 60 K	2.0
Sodium Carbonate	2.0
Water	80.0

Wax Stripper

Formula No. 1

(Nonionic-Anionic)

Water	84.00
Trisodium Phosphate	5.00
Sodium Metasilicate Pentahydrate	5.00
ESI-Terge HA-20	6.00
Ammonia or Monoethanolamine	3-5

Procedure:

Add in order listed with adequate agitation, allowing each material to dissolve before adding **ESI-Terge HA-20**. Agitate until clear.

Specifications:

Solids:	16%
Active:	16%
pH:	12.5-13.5
Visc.:	200 cps LV #2 spindle @ 60 rpm @ 25 C

No. 2

(Nonionic-Anionic)

Water	86.00
Trisodium Phosphate	3.00
Sodium Tripolyphosphate	3.00

ESI-Terge HA-20	5.00
Monoethanolamine	3.00

Procedure:
Add in order listed with adequate agitation, allowing each material to dissolve before adding **ESI-Terge HA-20**. Agitate 5 min.

Specifications:
Solids:	13.5%
Active:	13.5%
pH:	10.5-11.5
Visc.:	150 cps LV #2 spindle @ 60 rpm @ 25 C

No. 3

(Nonionic-Anionic, Liquid)

		Mixing Order
Bio Soft S-100	2.0	3
Monoethanolamine	5.0	2
Makon 10	1.0	5
Stepanate X	5.0	4
Trisodium Phosphate	5.0	8
Sodium Metasilicate	2.0	9
Urea	1.0	7
Sequestrene NA 4	1.0	6
Water	78.0	1
Fragrance and Colorant	q.s.	—

Procedure:
Combine 1, 2, and 3; mix until clear. Add balance of ingredients in sequence mixing thoroughly to dissolve each ingredient before adding next one.

pH:	11.9
Brookfield Visc.:	15 cps @ 25 C
Physical Appearance:	White opalescent liquid

No. 4

(Nonionic-Anionic)

Monamine ALX-100S	25.0
Monamine ADY-100	7.0
Water	68.0

Note:

This concentrate can be diluted up to 5:1 with water and then siphoned into a normal automatic car washing system.

No. 5

Condensate PN	10
Conco AAS 60 S	6
Sodium Tripolyphosphate	6
Tetrasodiumpyrophosphate	6
Conco SXS	7
Water	65

No. 6

(Amphoteric)

Miranol® JEM Conc.	4.0
Tetrapotassium Pyrophosphate	4.8
Trisodium Phosphate	3.0
Starso	5.0
Butyl Cellosolve	1.0
Water	82.2

pH (direct): 12.1
pH (1:10):   11.7
pH (1:20):   11.2

No. 7

(Amphoteric, Foaming)

Miranol® C2M-SF Conc.	5.0
Sodium Metasilicate Pentahydrate	5.0

| Tetrapotassium Pyrophosphate | 10.0 |
| Water | 80.0 |

Stripper–Cleaner–Degreaser

Formula No. 1

(Nonionic)

Butyl Cellosolve	53.00
*Potassium Hydroxide (45%)	8.86
ESI-Terge C-5	38.14

Procedure:
Add in order listed with adequate agitation.

Specifications of Concentrates:

Solids:	42.0
Active:	95.0
pH:	13-14
Visc.:	Water like

Specifications[1] of Cut Material:

Solids:	3.80
Active:	8.64
pH:	13-14
Visc.:	Water like

---

* Use proper protective clothing and safety equipment when handling potassium hydroxide.
[1] Cutting Instructions: 1 part concentrate to 10 parts water.

No. 2

(Anionic)

Water	87.00
*Potassium Hydroxide (90%)	2.40
TSP	0.60
Sodium Metasilicate	3.00
ESI-Terge 320	2.00
Butyl Cellosolve	5.00

Procedure:

Add in order listed with adequate agitation allowing each material to dissolve before adding the **ESI-Terge 320** and butyl Cellosolve.

Solids:	8.0%
Active:	13.0%
pH:	13.0-14.0
Visc.:	Water like

---

* Formulation Note: Up to 3% ammonia or 3% monoethanolamine may be added for additional strength. Proper protective clothing should be used when handling potassium hydroxide.

No. 3

(Nonionic-Anionic)

Water	87.00
*Potassium Hydroxide (90%)	2.40
TSP	0.60
Sodium Metasilicate Pentahydrate	3.00
ESI-Terge HA-20	2.00
Butyl Cellosolve	5.00

Procedure:

Add in order listed with adequate agitation, allowing each material to dissolve before adding the **ESI-Terge HA-20** and butyl Cellosolve.

Specifications:

Solids:	8.0%
Active:	13.0%
pH:	13.0-14.0
Visc.:	4 cps LV #1 spindle @ 60 rpm @ 25 C

---

* Formulation Note: Up to 3% ammonia or 3% monoethanolamine may be added for additional strength. Proper protective clothing should be used when handling potassium hydorxide.

Wax Stripper/Floor Cleaner

Formula No. 1

(Nonionic)

		Mixing Order
Ninol 1281	5	4
Sodium Metasilicate Pentahydrate	7	2
Tetrapotassium Pyrophosphate	3	3
Potassium Coco Soap (40%)	5	5
Water	80	1

pH:	12.1
Gardner Color:	2–
Brookfield Visc.:	About 400 cps @ 25 C
Appearance:	Clear light yellow liquid

No. 2

		Mixing Order
Stepan HDA-7	8.00	5
Morpholine	7.00	4
Trisodium Phosphate	2.75	3
Sodium Tripolyphosphate	2.25	2
Water	80.00	1

Procedure:
Dissolve phosphates in water. Add **HDA-7** and morpholine. Mix until uniform.

pH:	11
Gardner Color:	4+
Brookfield Visc.:	About 7 cps @ 25 C
Appearance:	Clear amber liquid

No. 3

(Nonionic)

Neodol® 91-6	4
Butyl Dioxitol®	4
Potassium Pyrophosphate	6
Water, Dye, etc.	q.s.

Properties:
Clear Point, C	0
F	32
pH, as is	10.1
Density, g/ml	1.048

Recommended dilution: 1/10 to 1/30

Floor Cleaner

Formula No. 1

(Anionic, Heavy Duty)

Trisodium Phosphate	16.75
Sodium Tripolyphosphate	30.00
Sodium Sesquicarbonate	51.9
Sulframin® 85 Flakes (1% LAS)	1.25
Perfume	0.10

No. 2

(Anionic)

Monamine ALX-100S	15.0
Water	85.0

Note:
Add perfume and, if desired, a disinfectant such as **Dowicide A** or **B**. Use one cupful in a bucket of water. For increased cleaning power, fortify with approximately 5% trisodium phosphate. If this product clouds at low temperatures, reduce the TSP or add blending agents such as sodium xylene sulfonate. For wax removal use 2 cups in a bucket of warm water.

No. 3

(Nonionic)

		Mixing Order
Ninol 1285	7	4
Trisodium Phosphate	5	3
Sodium Tripolyphosphate	5	2
Water	83	1

Procedure:
Combine ingredients and mix until uniform.

pH:	11.2
Gardner Color:	1+
Brookfield Visc.:	About 135 cps @ 25 C
Appearance:	Pale yellow liquid
Cloud Point:	Below 40 F

No. 4

(Nonionic)

		Mixing Order
Ninol 1285	6	3
Tetrapotassium Pyrophosphate	7	2
Water	87	1

Procedure:
Dissolve TKPP in water. Add **Ninol 1285.** Mix.

pH:	10.2
Gardner Color:	1−
Brookfield Visc.:	About 185 cps @ 25 C
Appearance:	Pale yellow liquid

No. 5

(Nonionic-Anionic, High Phosphate)

		Mixing Order
Stepanate X	19.0	3
Ninol 1285	8.6	4
Tetrapotassium Pyrophosphate	14.3	2
Butyl Cellosolve	9.5	5
Water	48.6	1

Procedure:

Dissolve the phosphate in the water. Add the other ingredients in the order listed with stirring.

pH:	10.4
Gardner Color:	1
Brookfield Visc.:	About 7 cps @ 25 C
Appearance:	Clear and bright, light yellow liquid

Garage Floor Cleaner (Concrete)

Formula No. 1

(Anionic)

Trisodium Phosphate (cryst.)	40
Sodium Tripolyphosphate	25
Soda Ash	25
Sulframin® LX or **85 Flake**	10

No. 2

(Nonionic)

Neodol 91-6	5
Tallow Soap	2
Butyl Oxitol®	6.5
Trisodium Phosphate	3
Sodium Metasilicate $5H_2O$	3
Water	q.s.

Procedure:

Stir **Neodol 91-6** and **Butyl Oxitol** to homogeneous mixture, add tallow soap and water. When the soap is dissolved, add trisodium phosphate and sodium metasilicate. The garage floor cleaner is applied directly to oil spots. After allowing 5 min for penetration, flush with garden hose.

Machine Floor-Scrubbing Cleaner

(Nonionic)

With **Antarox® BL-225** surfactant, liquid low-foaming, alkaline-built detergents can be formulated with sulfonate-type hydrotropes, such as sodium xylene sulfonate, for use in automatic floor cleaning equipment. The following typical formulation is clear and stable up to 50 C (122 F).

Antarox® BL-225	1.0
Tetrapotassium Pyrophosphate or Trisodium Phosphate	10.0
Sodium Xylene Sulfonate	6.6
Water	82.4

Bar Cleaner

Formula No. 1

(Nonionic)

		Mixing Order
Makon 10	10	2
Water	90	1

Procedure:

Combine ingredients and mix.

pH:	7.2
Brookfield Visc.:	About 10 cps @ 25 C
Appearance:	Colorless liquid

No. 2

(Nonionic)

		Mixing Order
Makon 10	5	1
Ninol 128	5	2
Water	90	3

Procedure:
Combine ingredients and mix until uniform.

pH:	10.1
Gardner Color:	1–
Brookfield Visc.:	About 100 cps @ 25 C
Appearance:	Pale yellow liquid

Brass, Magnesium, Steel, Lead Cleaner

(Anionic)

Nacconol 40F	20
Tetrasodium Ethylenediaminetetraacetate (dihydrate)	12
Sodium Carbonate	25
Monosodium Phosphate (monohydrate)	10
Sodium Metasilicate (pentahydrate)	33

Steel Tank Cleaner

(Anionic)

Nacconol 40F	5
Caustic Soda (flakes)	40
Sodium Carbonate	30
Tetrasodium Ethylenediaminetetraacetate (dihydrate)	15
Sodium Metasilicate (pentahydrate)	10

Metal Cleaner

(Anionic)

Formula	No. 1	No. 2	No. 3	No. 4	No. 5	No. 6	No. 7	No. 8	No. 9	No. 10	No. 11	No. 12
Neodol® 25-7	5	—	—	5	—	—	—	—	5	—	—	—
Neodol 25-9	—	5	—	—	5	—	5	—	—	5	—	10
Neodol 25-12	—	—	5	—	—	5	—	5	—	—	5	—
Sodium Metasilicate 5H$_2$O	1	1	1	2	2	2	2	2	2	2	2	1
Trisodium Phosphate	—	—	—	3	3	3	3	3	3	3	3	3
Tetrapotassium Pyro-phosphate	10	10	10	5	5	5	5	5	10	10	10	5
Butyl Oxitol®	—	—	—	—	—	—	5	—	—	—	—	—
Sodium Xylene Sulfonate	4	5	6	4	4	5	4	5	7	7	9	7
Water	q.s.[1]											
Solids (%w)	20	21	21	19	19	20	19	20	27	27	29	27
pH Concentrate	12.3	12.3	12.3	12.3	12.3	12.3	12.4	12.4	12.5	12.5	12.5	12.4
Clear Point (°C)	0	0	-2	1	-0.5	0	-2.5	0.5	2	-3	-2	-2
Cloud Point (°C)	54	62	69	67	66	77	54	90	84	84	93	70
pH[2] Use	10.3	10.3	10.3	10.5	10.5	10.5	10.4	10.4	10.6	10.6	10.6	10.4
Ross Miles Foam,[2] cm												
Initial	4.5	8.5	10	5.5	8.5	9	7	9	5.5	8	9	10
After 5 min	0.4	5.5	6.5	1.5	3	5.5	0.5	1.5	2	5	5	0.8

[1] Quantity sufficient to total 100%.
[2] Tested at use concentration (100 to 1 dilution of concentrate).

No. 13

(Nonionic-Anionic, Light Duty)

Plurafac RA-20 Surfactant	5
Klearfac AA-040 Surfactant	3
Na_2SiO_3	8
EDTA	15
Water	69

Note:
Use concentration: 2-4 oz/gal of water.

No. 14

(Cationic)

Kerosene	37
Arquad® 2C-75	2
Ethomeen® S/12	1
Phosphoric Acid Sol'n. (20%)	60

Procedure:
Dissolve the **Arquad 2C** and **Ethomeen S/12** in the kerosene and add to the phosphoric acid solution with vigorous agitation.

No. 15

(Cationic, Acid-Cleaning)

Phosphoric Acid (75%)	30
Monazoline O	5
Kerosene	5
Water	60

Truck Body Cleaner

(Anionic)

Nacconol 40F	50
Sodium Metasilicate (pentahydrate)	25
Sodium Tripolyphosphate (anhydrous)	25

Aluminum Cleaner

Formula No. 1

(Amphoteric, Acidic)

Miranol® C2M-SF Conc.	5.0
Butyl Cellosolve	6.0
Phosphoric Acid (85%)	38.0
Hydrofluoric Acid (70%)	8.0
Ethylenediamine Tetra-Acetic Acid	1.0
Water	42.0

No. 2

(Amphoteric)

Miranol® C2M-SF Conc.	5.0– 2.5
Butyl Cellosolve	6.0– 6.0
Phosphoric Acid (85%)	38.0–38.0
Hydrofluoric Acid (52%)	8.0– 8,0
Ethylenediamine Tetra-Acetic Acid	1.0– 1.0
Water	42.0–44.5

No. 3

(Amphoteric, Alkaline)

Miranol® C2M-SF Conc.	5.0–10.0
Sodium Metasilicate Pentahydrate	5.0– 5.0
Water	90.0–85.0

Note:

In nonindustrial aluminum cleaning operations it is not always possible to rinse immediately following the cleaning step. Should the parts dry between cleaning and rinsing a white film may appear, especially if silicates are used. This problem will not arise using the following formulation.

No. 4

(Amphoteric)

Miranol® C2M-SF Conc.	3.5
Tetrapotassium Pyrophosphate	5.0
Sodium Metasilicate Pentahydrate	1.0

Triton X-100	1.5
Carbitol	3.0
Water	86.0

pH: 1% sol'n.: 10.4
pH: direct: 11.2

Detergent Sanitizer

Formula No. 1

(Nonionic)

Biopal VRO-20	8.75
Phosphoric Acid (75%)	0.60
Dibutylthiourea	0.01
Igepal CO-880	5.00
Water	85.64

Dilution for hospital use:	use conc. in ppm iodine	av. oz per 2½ gal water
Nonporous hard surfaces	75–125	1½ to 2½
Porous surfaces	150	3

Note:

Formula 1 contains sufficient phosphoric acid to maintain a pH of 4.5 or lower after having been diluted to 75 ppm available iodine, provided that one uses relatively nonalkaline water, such as New York City tap water; additional acid will be required when this formulation is to be diluted with alkaline water. Formula 1 also contains sufficient dibutyl-thiourea to inhibit corrosion of galvanized iron equipment by the concentration of acid present at 75 ppm.

No. 2

(Nonionic)

Biopal VRO-20	4.00
Phosphoric Acid (75%)	4.00
Igepal CO-730	5.00
Water	87.00

Dilution:	use conc. in ppm iodine	av. oz per 10 gal water
Nonporous hard surfaces:	25-50	4¼-8½
Porous surfaces:	50-100	8½-16¾

Note:
This particular rinse is more suitable for applications in which a foam blanket is desired.

No. 3

(Amphoteric)

Miranol® C2M-SF Conc.	15.0
Quaternary Ammonium Salt Germicide (50%)	2.0
Sodium Carbonate	2.0
Ethylenediamine Tetra-Acetic Acid	0.5
Water	80.5

Disinfectant

Formula No. 1

(Chloroxylenol Solution)

Chloroxylenol	5.00
Terpineol	10.00
Alcohol (95%)	20.00
Castor Oil	6.30
Potassium Hydroxide	1.36
Crodolene	0.75
Water (deionized)	56.59

Procedure:
Dissolve the potassium hydroxide in 15 ml of purified water, add a solution of the **Castor Oil** in 63 ml of the alcohol, mix and allow to stand for an hour or until a small portion remains clear when diluted with nineteen times its volume of water. Then add the **Crodolene**, dissolve the PCMX in the remainder of the alcohol, mix with the **Terpineol** and add to the soap solution, finally adding sufficient deionized water to produce 1000 ml.

	No. 2	No. 3	No. 4
DCMX	2.00	2.00	2.00
Pine Oil	4.00	–	–
Terpinolene B	–	5.00	5.00
Ethyl Alcohol	–	20.00	–
Castor Oil Soap (30%)	12.00	20.00	34.00
Water (deionized)	2.00	53.00	59.00

No. 5

(Benzyl Cresol)

Benzyl Cresol	4.00
Castor Oil Soap (30%)	19.00
Terpineol BP	10.00
Ethyl Alcohol (IMS)	10.00
Water (deionized)	57.00

No. 6

(Aerosol Telephone Cleaner)

Irgasan DP 300	0.20
Dichlorophene	0.20
Crodamol OP	5.00
Isopropyl Alcohol	94.60

Pack: 25% concentrate
75% Propellant 11/12 (1:1)

Dairy Cleaner

Formula No. 1

(Nonionic)

		Mixing Order
Makon 10	10	4
Phosphoric Acid	25	3

Gluconic Acid	25	2
Water	40	

Procedure:
Dissolve gluconic acid in the water. Add phosphoric acid and **Makon 10** and mix.

pH:	1.5
Brookfield Visc.:	About 70 cps @ 25 C
Appearance:	Clear, colorless liquid

No. 2

(Amphoteric, Acid)

Miranol® C2M-SF Conc. (adjusted to pH 7.0)	13.27
Triton X-100	6.63
Quaternary Ammonium Salt Germicide (50%)	13.00
Glycolic Acid	17.00
Water	50.10

Emulsion Cleaner

Formula No. 1

(Anionic, Engine Degreaser)

Kerosene	80.0
Monafax L-10	10.0
Monafax H-15	10.0

Note:
The above concentrate can be diluted with at least 50% water and still have good storage stability.

No. 2

(Anionic, Degreaser)

Water	30.0
Monamulse 653-C	5.0
Monafax H-15	5.0
Mineral Spirits	60.0

	No. 3	No. 4	No. 5	No. 6
		(Nonionic-Anionic)		
Alkasurf I.P.A.M.	20	10	10	10
Alkasurf LA-3	—	10	—	10
Alkamide CDE	—	—	10	—
Kerosene	80	80	80	60
Perchloroethylene	—	—	—	20

Note:

These concentrates should be further diluted by 4-5 volumes of kerosene before using; this product is applied to the soiled surface by rushing, dipping, spraying. The solubilized and dispersed soil is rinsed away with water, leaving the surface clean.

No. 7

(Cationic, O/W)

Monazoline O or **T**	8.20
Mineral Oil	37.50
Glacial Acetic Acid	.75
Tall Oil Fatty Acid	3.55
Water	50.00

Procedure:

Blend first four ingredients in order listed; mix until clear. Add blend to water slowly with continuous agitation.

Note:

The blend of the first four ingredients is useful as a corrosion-inhibiting coating oil which is completely water rinsable.

No. 8

(Cationic, W/O)

Monazoline O or **T**	1.5
Mineral Oils	25.0
Tall Oil Fatty Acid	2.5
Water	71.0

Procedure:
Blend first three ingredients and follow with slow addition of water with continuous agitation. Further additions of water will increase the viscosity without causing emulsion phase inversion.

No. 9

(Amphoteric)

Miranol® C2M-SF Conc.	25.0
Triethanolamine	3.0
Alfonic 1012-40	13.0
Mineral Spirits (odorless)	45.0
Butyl Cellosolve	14.0

Offshore Oil Well Rig Cleaner

(Anionic)

Water	25
Isopropanol	25
Monawet MO65-150	10
Monamulse 947	40

Procedure:
Add the ingredients in the above order with moderate agitation.

Use Concentration:
Aspirate at the rate of approximately 1-2 oz/gal.

Note:
It has the additional advantage of forming fast breaking emulsions at use dilutions, so the disposal wastes can be collected and the oil easily separated before returning the water portion to the sea.

Oil Well Foaming Agent

(Nonionic-Anionic)

		Mixing Order
Steol CS-460	42	1
Makon 14	25	3

Water	25	4
Alcohol (denatured)	8	2

Procedure:

Mix **Steol CS-460** and alcohol. Add the other ingredients and stir until uniform.

pH:	9.7
Gardner Color:	1–
Brookfield Visc.:	About 90 cps @ 25 C
Appearance:	Clear, off white liquid

Acid-Type Equipment Cleaner

(Amphoteric)

Miranol® J2M Conc.	1.8
Miranol® C2M-SF Conc.	0.2
Gluconic Acid	6.0
Phosphoric Acid (75%)	62.0
Water	30.0

Note:

To prevent an odor of hydrochloric acid from appearing in the product, heat the first three ingredients of the formulation for 30 min at 55-60 C, then add the water and the phosphoric acid.

Spray Tank Aluminum Cleaner

Formula No. 1

(Amphoteric, Acid-Type)

Sulfuric Acid (concentrated)	40
Monateric LF-100	5
Water	55

Suggested Usage Level:
    1–2 oz/gal

	No. 2 *(Amphoteric, Phosphate)*	No. 3 *(Amphoteric, Nonphosphate)*
STPP	24	—
Metso (anhydrous)	24	32
Soda Ash	24	32
Caustic Soda	24	32
Monateric LF-100	4	4

Note:

These formulas, when used in spray tanks at only 1 oz/gal at 115 F, were found to be as good detergents as the commonly used low foaming nonionic based detergents which require 150 F operating temperatures for satisfactory cleaning and low foaming properties.

Oil-Tank Degreaser

(Amphoteric)

Antaron FC-34	10
Sodium Metasilicate	15-20
Water	75-70

Liquid Oil Slick Remover

(Nonionic)

Plurafac D-25 Surfactant	15
Plurafac A-24 Surfactant	5
Naphtha	40
Perchloroethylene	40

Engine Cleaner

Formula No. 1

(Nonionic, Kerosene-Based)

Kerosene	79.8
Monamulse 653-C	20.2

Usage:
Apply by spray aspirator with steam or water. This is readily rinsable with cold water.

No. 2

(Nonionic–Anionic, Stoddard Solvent-Based)

Stoddard Solvent	81.0
Monamulse 653-C	13.0
Monamine 1-76	6.0

Usage:
Same as for kerosene-based engine cleaner above.

Electric Motor Cleaner

		% in Aerosol
Concentrate:		70.0
Isotron 113	100.0	
Propellant:		30.0
Isotron 12	100.0	

Procedure:
May be pressure filled or cold filled.

Note:
Noncorrosive, nonflammable cleaner for electric motors.

Directions for use:
Spray directly on area to be cleaned.

Warning:
Contents under pressure. Do not puncture. Exposure to heat or prolonged exposure to sun may cause bursting. Do not three into fire or incinerator. Keep from children.

Package:
Tinplate container with paint-type valve.

Radiator Cleaner

Formula No. 1

(Nonionic)

Metso	45
Nullapon (30%)	45
Antarox A-400	10

Procedure:

First empty the radiator. Then dissolve the ingredients in warm water and use 1/3 oz to gallon of radiator capacity. Run the engine for a few days. Repeat the treatment if necessary.

No. 2

(Nonionic)

A combination of 3-5 parts by weight of **Plurafac A-24** with 95-97 parts by weight of oxalic acid will effectively clean automotive radiators. It removes rust, suspends scale and prevents the cooling system from clogging. Unlike anionic surfactants, **Plurafac A-24** is a low foamer.

Cleaner for Auto Radiators and Cooling System

(U.S. Patent 2,036,848)

Kerosene	4
Ortho-dichlorbenzol	7
Oleic Acid	1

Procedure:

Add the above to water (while circulating). This deposits on and softens dirt and grease. Then add 2 oz caustic soda and circulate to saponify grease and oil. Wash out with water.

Acidic Cleaner for Aluminum Trucks and Trailers

(Amphoteric)

Miranol® JS Conc.	10-10.0
Phosphoric Acid (75%)	50-50.0

Butyl Cellosolve	10-10.0
Ammonium Bifluoride	1- 3.0
Water	29-27.0

Note:
If high foam is required **Miranol® CS Conc.** should be used. Also, **Miranol® JS Conc.** has good acid inhibiting properties and can be used in pickling solutions.

Chain Lubricant

Formula No. 1

(Amphoteric)

Miranol® JEM Conc.	20.0
Ethylenediamine Tetra-Acetic Acid	5.0
Sodium Hydroxide (50%)	2.0
Water	73.0

No. 2

| Miranol® JEM Conc. | 20.0 |
| Water | 80.0 |

No. 3

Caprylic Acid	} neutralized to pH 11.6	10.0
Oleic Acid		10.0
Miranol® JEM Conc.		10.0
Water		70.0

Aircraft Cleaners

	No. 1	No. 2
	(Anionic)	
Neodol 25-3S (60% AM)	33.4	33.4
Sodium Xylene Sulfonate (40% AM)	—	5
Shell Cyclo Sol® 63[1]	45.85	45.85

Shell Sol® 340[2]	9.75	9.75
Butyl Oxitol	10	5
Sodium Nitrite	1	1

---

[1] Aromatic hydrocarbon; flash point 142 F/61 C, b.p. 358–410 F/181–211 C.
[2] Aliphatic naphtha; flash point 104 F/40 C, b.p. 316–358 F/158–181 C.

Commercial Car Wash

Formula No. 1

(Nonionic, Solid Concentrate)

Neodol 23-6.5	20
Sodium Carbonate	40
Sodium Pyrophosphate	40

No. 2

(Nonionic, Liquid Concentrate)

Neodol 91-6 or **Neodol 23-6.5**	68
Butyl Oxitol®	28
Potassium Hydroxide	0.5
Water	3.5

No. 3

(Nonionic–Anionic, High Viscosity Liquid)

Neodol 23-6.5	15
Neodol 25-3A (60% AM)	5
Hydroxyethylcellulose[2]	1
Water	q.s.[1]

---

[1] Quantity sufficient to make 100% total.
[2] **Natrosol 250 HR.**

	No. 4	No. 5	No. 6	No. 7
		(Nonionic–Anionic)		
Neodol® 91-6	15	15	15	15
Neodol 25-3S (60% AM)	–	–	5	–
LAS (35% AM)	5	–	–	–
Potassium Pyrophosphate	5	–	5	5
Water	q.s.[1]	q.s.[1]	q.s.[1]	q.s.[1]

---

[1] Quantity sufficient to make 100% total.

[2] Dissolve **Neodol 91-6** and anionic surfactant in water first. Then add pyrophosphate.

Tire Cleaner

(Amphoteric)

An aqueous spray containing 15% **Antaron FC-34** and 10–15% sodium metasilicate can safely be used for cleaning white-wall tires. This solution is harmless to the metal and finish of the car as well as to the white wall itself. Application is preferably by spraying all the tires and subsequently rinsing them off in the same order.

Commercial Spray-Vac Rug Cleaner

(Amphoteric–Anionic)

Water	78.35
TKPP	3.35
EDTA, Na$_4$ (40% sol'n.)	0.50
Monateric LF Na-50	12.80
Monawet SNO-35	5.00

pH (as is):   approx. 11.5
pH (use conc.):   approx. 9.6

Suggested ussage level:
    2 oz/gal

Note:
    Monateric LF Na-50 is preferred over **Monateric LF-100** because the

former is more quickly dissolved in water.

Monawet SNO-35 is incorporated because of its unique wetting properties. Above 140 F it is an excellent wetting agent, but at lower temperatures it is a poor wetting agent. Thus, at application temperatures it provides excellent initial wetting or "strike," but on cooling to room temperature it loses its wetting character which permits the rug to dry faster.

Institutional Hand Dishwash and Glass Cleaner

(Anionic)

Monamine ALX-100S	18.0
Triton X-100	12.0
Water	70.0

Terrazzo Cleaner

(Anionic)

Monamine ALX-100S	8.0
Monaquest CA-100	2.0
Water	90.0

Directions for use:
Use 1–3 cupfuls in 3 gal of water according to performance requirements.

Industrial Wall Cleaners

Formula No. 1

(Nonionic, Spray-Type)

Neodol® 23-6.5	2.4
Tetrasodium Ethylenediaminetetraacetate	2.6
Butyl Oxitol®	3.0
Isopropyl Alcohol (99%)	1.0
Water, Dye, Perfume	q.s.[1]

---

[1] Quantity sufficient to make 100% total.

No. 2

(Nonionic, Spray-Type)

Neodol 23-6.5 or **Neodol 25-3S** (60% AM)	1.7
Lauric Diethanolamide	0.5
Sodium Metasilicate 5H$_2$O	1.7
Trisodium Phosphate	1.0
Butyl Oxitol®	3.5
Water, Dye, Perfume	q.s.[1]

[1] Quantity sufficient to make 100% total.

No. 3

(Nonionic, Liquid Light Duty)

Neodol 91-6 or **Neodol 91-8**	5
Trisodium Phosphate	2
Sodium Metasilicate 5H$_2$O	2
Water	q.s.[1]

[1] Quantity sufficient to make 100% total.

Mechanics' Hand Cleaner

(Fast Acting for Heavy Soils)

Stoddard Solvent	22.9
Solvesso 150	25.9
Monamulse 653-C	12.8
Water	38.4

Institutional Liquid Machine Dishwashing Compound

(Amphoteric)

Miranol® JEM Conc.	3.0
Tetrapotassium Pyrophosphate	10.0
Sodium Metasilicate Pentahydrate	10.0
Water	77.0

Note:

Safe for use on aluminum pots and pans.

Industrial Laundry Detergent

Formula No. 1

(Anionic)

Sodium Metasilicate (anhydrous)	87
ESI-Terge 320	4
Optical Brightener	0.1-0.3
Sodium Tripolyphosphate (anhydrous)	9

Procedure:

To a suitable powder mixer add sodium metasilicate and **ESI-Terge 320.** Allow the mix 10 min and add the brightener and sodium tripolyphosphate. Allow an additional 10 min of mixing.

	No. 2	No. 3	No. 4	No. 5
		(Nonionic)		
A **Neodol® 23-6.5**	20	10	–	–
Neodol 91-6	–	10	–	–
Neodol 91-8	–	–	10	–
Neodol 25-9	–	–	10	20
CMC	1	1	1	1
Water	79	79	79	79
Cloud Point, °C, 1% sol'n.	41	46	76	73
Visc., cps/21 C	1007	203	94	122
B Sodium Silicate[1]				
(2.4:1 SiO_2:Na_2O)	5	5	5	5
Potassium Hydroxide	14	14	14	14
Tetrapotassium Pyrophos-				
phate	10	10	10	10
Water	71	71	71	71

[1] Philadelpha Quartz, RU grade, 46%.

Procedure:
Blend parts A and B separately and deliver to wash water in the ratio
A:B = 1:4.

No. 6

Sodium Alkyl Benzene Sulfonate (**Nansa HS85/S**)	5-7 A.I.
Empicol TAS	5-7 A.I.
Empilan CME	2-3
Empiphos STP	35.0
Sodium Silicate	≈ 6.0
Sodium Perborate	≈ 20.0
Optical Brightening Agent	0.5
Sodium Carboxymethyl Cellulose	1.5
Enzyme	0.005-0.01 Anson units/g
Perfume, Color, Fillers, etc.	balance

No. 7

(Controlled Foam)

Sodium Alkyl Benzene Sulfonate	2- 5
Empicol TAS	5- 2
Soap (tallow)	5- 8
Empilan KM11	0- 5
Empiphos STP	35-40
Sodium Silicate	≈ 6
Sodium Perborate	≈ 20
Optical Brightening Agent	0.5
Sodium Carboxymethyl Cellulose	1.5
Perfume, Soap, Antioxidant, Fillers	balance

Wool Scouring

Formula No. 1

(Anionic)

Nacconol 40F	2.5
Sodium Bicarbonate or Sodium Chloride	8.4
Water	(100 gal) 835

Sufficient for at least 40 lb of wool (5-6% mineral oil).

No. 2

(Anionic)

Nacconol 40F	1.7
Sodium Bicarbonate or Sodium Chloride	8.4
Water	(100 gal) 835

Sufficient for at least 40 lb of wool (5-6% mineral oil).

Printing Roller Cleaner

(Cationic)

Kerosene	97
Nitropropane	2
Monazoline O	1

Industrial Window Cleaners

Formula No. 1

(Anionic, Windex-Type)

Neodol 25-3A (60% AM)	0.15
Isopropyl Alcohol	5
Ammonia (conc.)	0.15
Water, Dyes	q.s.

	No. 2	No. 3	No. 4
		(Nonionic-Anionic)	
Neodol® 25-3S (60% AM)	0.1	—	0.15
Neodol 23-6.5	—	0.1	—
Potassium Pyrophosphate	—	—	0.02
Butyl Oxitol®	2.5	—	0.1

Methyl Oxitol	2.0	–	–
Isopropyl Alcohol	20	15	–
Water (deionized), Dyes, etc.	q.s.[1]	q.s.[1]	q.s.[1]

---

[1] Quantity sufficient to make 100% total.

Chapter V

SOLVENT CLEANERS

Paint Cleaner

(Anionic)

Conco AAS 35 S	12.0
Tetrapotassium Pyrophosphate	1.0
Carbitol Solvent	4.0
Water	83.0

Paint Stripper

Formula No. 1

(Anionic)

A	**Veegum T**	0.75
	Kelzan	0.25
	Water	62.00
B	**Ultrawet 40 SX**	1.00
	Gafac RE-610	2.50
C	Sodium Hydroxide	15.00
	Water	18.50

Procedure:

Dry blend the **Veegum T** and **Kelzan** and add to the water slowly, agitating continually until smooth. Add part B to A. Combine C and add to A and B. Mix until uniform.

Directions for use:

Apply liberally with a brush to painted metal surface. Allow to stand until old finish is loosened from surface (10-20 min). Remove old finish with scraper or steel wool. Rinse surface with water.

Caution:
   Contains caustic. Wear rubber gloves.

No. 2

Methylene Chloride	67.00
Methanol	14.00
Celacol MM1PR1	2.00
Ethylene Glycol Mono Acetate	8.00
Paraffin Wax	1.00
Toluol	6.00
Wetting Agent	1.00-3.00
Corrosion Inhibitor	q.s.
Activator	1.00

No. 3

Methylene Chloride	80.00
Methanol	15.00
Celacol MM1PR1 or **MMP20**	1.80
Paraffin Wax	1.50
Wetting Agent	1.00
Activator	1.00
Corrosion Inhibitor	q.s.

No. 4

A **Veegum PRO**	1
Klucel M	1
B Water	25
N-Methyl-2-Pyrrolidone	73

Procedure:
   Dry blend the **Veegum PRO** and **Klucel** and add to B slowly, agitating continually until smooth.

Directions for use:
Apply evenly with brush to painted surface. Allow to stand for 10-30 min. Remove old finish with metal scraper or steel wool. Apply additional coats if necessary. Wipe wood surface with turpentine or denatured alcohol.

No. 5

(Anionic, Industrial Solid Alkaline)

Sodium Hydroxide	85
Sodium Lignosulfonate	6
Sodium Heptonate Dihydrate	5
Cresylic Acid	3
Anionic Surfactant	1

Use at a rate of 1-2 lb/gal at 90 C.

No. 6

(Industrial Liquid Alkaline)

Sodium Hydroxide (100° Tw)	83
Sodium Heptonate (industrial)	12
Cresylic Acid	5

Use at a rate of 2-4 lb/gal above 90 C.

Note:
The excellent heat stability of sodium heptonate makes it particularly suitable for this application. Such formulations will cope with a wide variety of primers, epoxy resins and acrylics though not with the most sophisticated modern finishes without some degree of reformulation.

Glass Cleaner

Formula No. 1

(Anionic)

Monamine ALX-100S	18.0
Triton X-100	12.0
Water	70.0

No. 2

(Anionic)

Concentrate:		90.0
Dow Corning EF-1-0108 Emulsion	1.5	
Triton W-30	0.2	
Isopropanol (99%)	2.0	
Dowanol EE	0.3	
Water	96.0	
Propellant:		10.0
Propellant 114	60.0	
Propellant 12	40.0	

Procedure:

Mix all ingredients except silicone emulsion. With gentle stirring, add the mixture to the silicone emulsion.

Package:

Lacquer-lined cans with appropriate commercial valve and mechanical break-up botton.

Directions for use:

Spray on glass surface lightly. Wipe dry with a soft clean cloth.

Warning:

Contents under pressure. Do not puncture. Exposure to heat or prolonged exposure to sun may cause bursting. Do not throw into fire or incinerator. Keep from children.

No. 3

(Nonionic-Anionic, Ultrasonic)

Nacconol 90F	6
Surfynol 485	5
Disodium Ethylenediaminetetraacetate (dihydrate)	12
Sodium Carbonate (anhydrous)	27
Sodium Dihydrogen Phosphate (monohydrate)	12
Sodium Metasilicate (pentahydrate)	38

Note:

It is suggested that this formulation be used as a 7% (by weight) aqueous solution at 180 F or higher.

Rinse Aid

	Formula No. 1*	No. 2*	No. 3**	No. 4**
		(Nonionic)		
Antarox® BL-240	70	70.0	50	50
Isopropanol	—	—	16	9
Urea	—	4.5	—	—
Water	30	25.5	34	41

---

* 70% Active
** 50% Active

Note:

It is also possible to prepare very low cost rinse aid concentrates by merely diluting **Antarox BL-240** with water; a 70% concentrate has a cloud point of 50 C.

	No. 5	No. 6	No. 7
		(Nonionic)	
Antarox® BL-240	20.0	20.0	20.0
Igepal® CO-430	20.0	20.0	20.0
Hydroxyacetic Acid (70%)	21.3	21.3	21.3
Isopropanol	10.0	10.5	11.0
Water	28.7	28.2	27.7

Window Cleaner

Formula No. 1

(Anionic)

Sulframin® 45 Liquid	10
Isopropanol	30
Water	60

No. 2

(Anionic, For Direct Application)

		Mixing Order
Nacconol 90F	0.1	2
Water	58.9	1
Diethylene Glycol Monoethyl Ether	6.0	3
Isopropyl Alcohol	35.0	4
Colorant and Perfume	q.s.	5

Procedure:
Combine ingredients in order listed.

No. 3

(Nonionic, Aerosol)

Crillon LDE	0.20
Volpo NP9	0.40
Isopropyl Alcohol	5.00
Water (deionized)	94.00

Fill: 95% product
       5% Butane

No. 4

(Nonionic, Liquid Aerosol)

		Mixing Order
Makon 10	0.05	2
Stepantex WB-42	0.05	3
Silicone SF-1066	0.20	5
Isopropyl Alcohol	10.00	4
Water (deionized)	86.70	1
Ammonium Hydroxide (28%)	3.00	6

Procedure:
Combine ingredients in the order listed.

Aerosol:
The load should contain 96 parts by weight of the foregoing recipe and 4 parts by weight of isobutane.

No. 5

(Nonionic-Anionic)

Ultra Blend® 100	5
Isopropanol	30
Water	65

No. 6

(Aerosol)

Cropol 60	0.025-0.30
Isopropyl Alcohol	64.50
Water (deionized)	25.00
Propellant 12	10.00

No. 7

(Aerosol)

Silicone Oil (350 cs)	1.00
Crill 4	2.00
Deodorized Kerosene	10.00
Crillet 4	2.00
Water (deionized)	65.00
Isopropyl Alcohol	20.00

Pack: 95% concentrate
          5% propellant

No. 8

(Solvent-Type)

Isopropyl Alcohol	66.75
Water (deionized)	33.15
Crodasinic LS35	0.10-0.30

Procedure:
Blend all components.

No. 9

(Abosrbent-Type)

Volpo T10	3.00
Kaolin	17.00
Petroleum Spirit SBP6	40.00
Water (deionized)	40.00

Procedure:
Blend all components except water by stirring (no heat is required). Then add cold water slowly with agitation and stir until a smooth liquid emulsion forms.

No. 10

(Ammoniacal-Type)

Carbitol	4.00
Isopropyl Alcohol	16.00
Water (deionized)	77.50
Ammonia 880	2.00
Crillon LDE	0.50

Procedure:
Simply blend all components.

Note:
The ammonia content may be adjusted as required.

Windshield Washer

Formula No. 1

(Anionic)

Nacconol 90F	0.2
Water	49.8
Isopropyl Alcohol or Methyl Alcohol	50.0
Alphazurine 2G or Blue Dye	as desired

Recipe for washer jar (one quart) for winter use:
Ounces of formula	8
Water	fill jar

Recipe for washer jar (one quart) for summer use:
Ounces of formula	2
Water	fill jar

No. 2

(Anionic)

Isopropanol	39
Water	60
Sulframin® 90 Flakes	1

Procedure:
Dissolve **Sulframin® 90 Flakes** in water, then add alcohol.

No. 3

(Anionic)

		Mixing Order
Steol CA-460	1	2
Isopropyl Alcohol	9	1
Water	90	3

Procedure:
Combine alcohol and **Steol CA-460**. Stir until uniform. Add water and stir.

pH: 7.4   If necessary, adjust with citric acid to lower pH;
NaOH solution to increase pH.
Appearance: Clear, colorless, water thin liquid.

No. 4   No. 5

(Nonionic-Anionic, Liquid Concentrate)

	No. 4	No. 5	Mixing Order
Nacconol 90F	0.2	–	1
Makon 10	–	0.2	1'
Water	49.8	49.8	2
Isopropyl Alcohol or			
Methyl Alcohol	50.0	50.0	3
Alphazurine 2G Blue Dye	q.s.	q.s.	4
Perfume	q.s.	q.s.	5

Procedure:
Dissolve 1 or 1' in 2. Add balance of ingredients in the order listed.

For winter use:
Dilute 8 oz of concentrate with 24 oz of water.

For summer use:
Dilute 2 oz of concentrate with 30 oz of water.

Lens Cleaner

(Nonionic)

Isopropyl Alcohol	85.0
Triton X-100	0.1
Water	14.9

Procedure:
Mix all ingredients together until smooth.

Eyeglass Cleaner

Ammonium Soap	6.40 oz
Glycerin	6.60 oz
Water	q.s. 1.00 gal
Fluorescein, soluble (D & C Yellow No. 7)	sufficient

Procedure:

Dissolve the soap in the water, add the glycerin and tint it with the fluorescein. Shake well. Place a few drops of this preparation on the eyeglass and wipe off clean with a soft cloth.

Car Wash Detergent

Formula No. 1

(Anionic, Liquid)

Alkasurf LA Acid	20
Caustic Soda Flake	2.6
Alkaquest EDTA	5
Water	to 100
Alkasurf SX 40	20

Note:

Adjust pH to 7-8.

No. 2

(Anionic)

Alkasurf LA Acid	20
Caustic Soda Flake	2.6
Alkaquest EDTA	5
Water	to 100
Alkasurf SX 40	20

Note:

Adjust pH to 7-8.

No. 3

(Anionic, Liquid Concentrate)

		Mixing Order
Nacconol 90F	17.0	4
Ethylenediaminetetraacetic Acid	5.0	3
Monoethanolamine	3.0	2
Stepanate X	5.0	5
Water	70.0	1
Colorant and Fragrance	optional	6

Procedure:

Combine ingredients in the order listed. Heat to 50 C to dissolve 4. Adjust pH to about 7.0 with monoethanolamine. Mix until clear. Perfume and color as desired.

pH:	7.0
Gardner Color:	2+
Brookfield Visc.:	140 cps @ 25 C
Appearance:	Clear, yellow liquid.

No. 4

(Anionic, Powder)

		Mixing Order
Nacconol 40F	20.0	5
Sodium Tripolyphosphate (light, granular)	47.0	1
Sodium Sesquicarbonate	25.0	2
Trisodium Phosphate	5.0	3
Sequestrene NA-4	3.0	4
Colorant and Fragrance	optional	6

Procedure:

Combine all ingredients except **Nacconol 40DBX**; blend until homogeneous. Add **Nacconol 40DBX**, mix 1-2 min and pack out via 10 mesh sieve.

Note:
For use in spray-machines.

Equipment:
Twin-shell blender, rotating drum, plowshare mixer, ribbon blender, or equivalent.

No. 5

(Anionic, Liquid)

Nacconol 35SL	60
Water	40

(Visc. can be controlled by addition of carboxymethylcellulose, an alkanolamide, or various vegetable gums.)

No. 6

(Anionic, Concentrate)

Water	84.50
Monaquest CA-100	1.00
Monamine 779	11.00
Monamine ALX-100S	3.50

Procedure:
Add ingredients in order listed and mix with slow agitation. Approximate visc. − 175 cps. pH (as is) − 9.1.

Usage:
This concentrate can be diluted up to 4:1 with water and then siphoned into a normal automatic car wash.

No. 7

(Nonionic, Powdered)

Sodium Tripolyphosphate (granular)	63
Alkyl Benzene Sulfonate Powder (40%)	30

Alkasurf LA-16	2
Alkamide CDO	5

No. 8

(Nonionic-Anionic, Liquid Superconcentrate)

		Mixing Order
Bio Soft N-300	40.0	1
Ninol 128 Extra	40.0	4
Amidox C-5	10.0	3
Propylene Glycol	10.0	2

Procedure:

Combine ingredients in the order listed. Adjust pH to 6.5-7.2 with dilute sulfuric acid (10%).

pH:	6.9
Gardner Color:	2+
Brookfield Visc.:	180 cps @ 25 C
Appearance:	Clear, yellow liquid

No. 9

(Nonionic-Anionic, Commercial, Liquid Concentrate)

		Mixing Order
Steol KS-460	75.0	2
Ninol 128 Extra	15.0	3
Water	10.0	1
Colorant and Fragrance	optional	4

Procedure:

Combine ingredients in the order listed, mixing 2-5 min after addition of each ingredient. Adjust pH to 7.0-7.5 with dilute sulfuric acid. Perfume and color as desired.

pH:	7.2
Gardner Color:	2+
Brookfield Visc.:	80 cps @ 25 C
Appearance:	Clear, yellow liquid

No. 10

(Nonionic-Anionic, Commercial, Liquid Concentrate)

		Mixing Order
Bio Soft N-300	73.0	2
Ninol 128 Extra	15.0	3
Propylene Glycol	2.0	4
Water	10.0	1
Colorant and Fragrance	optional	5

Procedure:

Combine ingredients in the order listed, mixing 2-5 min after addition of each ingredient. Perfume and color as desired.

pH:	7.7
Gardner Color:	3–
Brookfield Visc.:	825 cps @ 25 C
Appearance:	Clear, yellow liquid

No. 11

(Nonionic-Anionic, Concentrate)

		Mixing Order
Bio Soft S-100	34.58	3
NH$_4$OH (28%)	6.85	2
Makon 10	8.50	5
Ninol AA-62	6.20	4
SD 3A Alcohol	17.00	6
Water	26.87	1

Procedure:

Combine water and NH$_4$OH. Neutralize with **Bio Soft S-100**. Add melted (50-60 C) **Ninol AA-62 Extra** and stir. Add **Makon 10** and ethanol. Adjust pH.

      pH: 6.8-7.2   If necessary, adjust with sulfuric acid if too high; NH$_4$OH if too low.

Gardner Color:	1+
Brookfield Visc.:	About 35 cps @ 25 C
Appearance:	Clear, light yellow liquid

No. 12

(Nonionic-Anionic, Concentrate)

		Mixing Order
Bio Soft N-300	43.30	3
Steol CS-460	16.70	4
Ninol AA-62 Extra	4.00	2
Phosphoric Acid	0.25	5
Water	q.s.	1

Procedure:

Combine water and **Ninol AA-62 Extra**. Heat (60 C) with stirring until **Ninol AA-62 Extra** is melted. Add **Bio Soft N-300**. After the **Bio Soft N-300** is mixed in add the remaining ingredients and stir until uniform. Adjust pH. If necessary add ethanol to clear.

      pH: 7.5-7.8   If necessary, adjust with phosphoric acid if too high; NaOH if too low.

Gardner Color:	1-
Brookfield Visc.:	About 700 cps @ 25 C
Appearance:	Clear, pale yellow liquid

No. 13

(Nonionic-Anionic, Concentrate)

		Mixing Order
Bio Soft S-100	13.00	6
NaOH (50%)	3.00	4

Makon 10	13.50	5
Ninol 2012 Extra	20.00	7
Stepanate X	6.25	3
Urea	9.50	8
Propylene Glycol	1.00	2
Water	33.75	1

Procedure:
Combine water, NaOH, **Stepanate X**, and propylene glycol. Neutralize by adding **Bio Soft S-100** with stirring. Add **Ninol 2012 Extra**. Adjust pH. Add urea.

pH: 7.1-7.3   If necessary, adjust with sulfuric acid if too high; NaOH if too low.

Gardner Color: 1—
Brookfield Visc.:   About 500 cps @ 25 C
Appearance: Clear, pale yellow liquid

Can be diluted to give approximately the following viscosities:

Dilution (as is)	Approximate Visc. (cps)
20 %	500
17.5%	450
15 %	400

No. 14

(Nonionic-Anionic, Liquid)

Condensate CO	7.0
Conco AAS 35 S	2.0
Conco NI-100	4.0
Water	87.0

No. 15

(Nonionic-Anionic, Concentrate)

| Conco AAS 60 S | 60.0 |
| Condensate PS | 40.0 |

No. 16

(Nonionic-Anionic, Shampoo)

Alkasurf ES 60	30
Alkamide CME	2
Water	58
Color	q.s.

Procedure:

The **Alkamide CME** is melted and mixed with slightly warmed **Alkasurf ES 60** to which the water has already been added. To the clear solution, the color is blended as required.

No. 17

(Nonionic-Anionic, Concentrate)

Water	84.50
Monaquest CA-100	1.00
Monamine 779	11.00
Monamine ALX-100S	3.50

Procedure:

Add ingredients in order listed and mix with slow agitation. Approximate visc.: 175 cps. pH "as is": 9.1.

Usage:

This concentrate can be diluted up to 4:1 with water and then siphoned into a normal automatic car wash.

No. 18

(Nonionic-Anionic)

Alkasurf LA-Acid	5-10
Alkasurf LA-16	2-8
Alkamide CDO	2-4
Caustic Soda	1-2
Preservative	q.s.
Color	q.s.
Water	to 100

No. 19

(Amphoteric, Shampoo)

Miranol® C2M-SF Conc.	15.0
Sodium Lauryl Ether Sulfate (28%)	10.0
Water	75.0

Note:

Miranol® C2M-SF Conc. may be replaced by the salt-free linoleic derivative Miranol® L2M-SF Conc. or the salt-free oleic derivative Miranol® OM-SF Conc. If 1% water is replaced with sodium benzoate or sodium nitrate, this formulation may be packed in unlined containers. The viscosity may be increased by adding 3 to 5% of lauric diethanolamide.

No. 20

(Amphoteric, Shampoo)

Miranol® C2M-SF Conc.	15.0
Water	85.0

Note:

Miranol® C2M-SF Conc. may be replaced by the salt-free linoleic derivative (Miranol® L2M-SF Conc.) or the salt-free oleic derivative (Miranol® OM-SF Conc.). If 1% water is replaced with sodium benzoate or sodium nitrate, this formulation may be packed in unlined containers. The viscosity may be increased by adding 3 to 5% of an amine condensate.

No. 21

(Jet, Concentrate)

Water	70
Detergent Concentrate 840	30

Recommended Usage:

1/3 of an oz/gal of water. This dilution (approximately 1:400) produces excellent detergency and foam even when recycled through a recovery tank. The foam is stable in the presence of soil.

Improved cleaning properties can be achieved by adding 2% sodium metasilicate anhydrous to the above formulation. Also, 1% trisodium phosphate can be incorporated, if conditions allow its use. At 1-2 oz/gal use level, these built systems are effective truckbody cleaners and cosmolene detergents.

No. 22

(Cationic)

Carnauba Spray 200	25
Kerosene or Light Mineral Oils	25
Water	50

Note:
  The shipping container should be thoroughly agitated before formulating is begun at ambient temperature. Tap water and a good grade of kerosene or light mineral oil may be used. A heresite or equivalent protective lining should be used in the formulation tank to prevent vapor phase rusting.
  It is recommended that 10 gal of wax formulation be added all at once to 40 gal of water in a drum at the car wash station. A clear, uniform solution will result. The application rate should be 2-6 oz of the diluted end product per auto during the final rinse.

Appearance:	Semiliquid, layered
Color:	Light amber
Solvents Present:	Water, isopropyl alcohol
Approximate Concentration (%):	68
Specific Gravity (20/20 C):	0.880
Weight/Gallon, pounds:	7.3
Shipping Container and Weight:	400 lb tighthead steel drum

No. 23

Carspray #2 Concentrate	25
Kerosene	25
Water	50

Note:
   Tap water and a good grade of kerosene should be used to make the dispersion at ambient temperature. A few percent isopropyl alcohol may be added to eliminate haziness. Water soluble dye may be added to color the product. A heresite or equivalent protective lining should be used in the formulation tank to prevent vapor phase rusting.
   It is recommended that 10 gal of rinse aid formulation be added to 40 gal of water in a 55 gal drum at the car wash station. The application rate should be 2-6 oz of the diluted end product per auto during the final rinse.

Appearance:	Clear liquid
Color:	Yellow
Solvents Present:	Water, Butyl Cellosolve
Approximate Concentration (%):	60
Specific Gravity (20/20 C):	0.870
Weight/Galllon, pounds:	7.2
Shipping Container and Weight:	400 lb tighthead steel drum

No. 24

Carspray 300	15-25
Coupling Solvent	3-9
Kerosene or Mineral Oils	≈ 25
Water	≈ 50

Note:
   The characteristics desired in the end product will determine composition. The ingredient levels shown here are merely suggested useful ranges.
   The formulator should determine dilution recommendations for the car wash operator. The diluted end product should be effective at use levels of 2-6 oz per auto during the final rinse.

Appearance:	Clear liquid
Color:	Yellow
Solvents Present:	Isopropyl alcohol
Approximate Concentration (%):	75
Specific Gravity (20/20 C):	0.876
Weight/Gallon, Pounds:	7.3
Shipping Container and Weight:	390 lb tighthead steel drum

No. 25

Carspray 436	15-25
Coupling Solvent	3-9
Kerosene or Mineral Oils	≈ 25
Water	≈ 50

Note:

The characteristics desired in the end product will determine composition. The ingredient levels shown here are merely suggested useful ranges.

The formulator should determine dilution recommendations for the car wash operator. The diluted end product should be effective at use levels of 2-6 oz per auto during the final rinse.

Appearance:	Clear liquid
Color:	Pale yellow
Solvents Present:	Water, isopropyl alcohol
Approximate Concentration (%):	75
Specific Gravity (20/20 C):	0.876
Weight/Gallon, Pounds:	7.3
Shipping Container and Weight:	390 lb tighthead steel drum

Whitewall Tire Cleaner

Formula No. 1

(Anionic)

Monamine ALX-100S	15.0
Monaquest CA-100	5.0
Water	80.0

Note:

This cleaner should be used without further dilution and then flushed off with generous amounts of water.

No. 2

(Anionic)

		Mixing Order
Stepan HDA-7	8	5
Trisodium Phosphate	6	4

Sodium Tripolyphosphate	5	3
Stepanate X	5	2
Water	76	1

Procedure:

Dissolve the phosphates in the water and **Stepanate X**. Add the **Stepan HDA-7** and stir until uniform.

pH:	11.4
Gardner Color:	5
Brookfield Visc.:	About 6 cps @ 25 C
Appearance:	Clear, light brown liquid

No. 3

(Anionic)

		Mixing Order
Ninex 24	7	4
Sodium Metasilicate Pentahydrate	10	3
Potassium Hydroxide	3	2
Water	80	1

Procedure:

Dissolve KOH and silicate in the water. Add **Ninex 24** and stir until homogeneous.

pH:	12.4
Gardner Color:	1−
Brookfield Visc.:	About 5 cps @ 25 C
Appearance:	Clear, pale yellow liquid

Carbitol or isopropanol can be added.

No. 4

(Anionic, Liquid)

| **Nacconol 90F** | 2.0 |
| Trisodium Phosphate (monohydrate) | 5.0 |

Methocel 1500	0.5
Sodium Metasilicate (pentahydrate)	5.0
Water and Colorant	87.5

No. 5

(Anionic, Powder)

Nacconol 90F	25
Sodium Metasilicate (pentahydrate)	20
Trisodium Phosphate (dodecahydrate)	15
Sodium Xylene Sulfonate (powder)	10
Sodium Carbonate (anhydrous)	30

Directions for use:
Apply as aqueous paste to the white walls and after 2-5 min contact, flush off with generous amounts of water.

No. 6

(Anionic)

Nacconol 35SL
Trisodium Phosphate (hydrate)
Tetrasodium Pyrophosphate
Water, Dye, Perfume, Preservative

No. 7

(Nonionic, Liquid)

Water	73
Sodium Tripolyphosphate	5
NaOH (50%)	10
Sodium Silicate (42°, 3.22/1)	8
Alkali Surfactant	2
Triton X-100	2

Wt./gal: 9.5 lb
Chelating power: adequate for 25 to 1 dilution in 400 ppm hard water.

Stability: stable from 35-120 F, usable to 212 F
pH at dilution:

as is	13.2
5 to 1	13.0
10 to 1	12.7
20 to 1	12.5
50 to 1	12.0

No. 8

(Nonionic, Alkaline)

		% in Aerosol
Concentrate:		85.0
$Na_3PO_4 \cdot 12H_2O$	10.0	
Igepal CO-630	5.0	
Water	85.0	
Propellant:		15.0
Isotron 12	10.0	
Isotron 114	90.0	

Procedure:
Dissolve the phosphate and **Igepal** in the water and pressure fill.

Package:
Seamless tinplate container with a mechanical breakup spray valve or a foam valve.

Directions for use:
Shake well before using. Spray onto tire, let stand for several minutes, then rinse thoroughly. Stubborn stains may require a second application.

Warning:
Contents under pressure. Do not puncture. Exposure to heat or prolonged exposure to sun may cause bursting. Do not throw into fire or incinerator. Keep from children.
Do not allow to remain on skin. Do not spray near eyes.

No. 9

(Nonionic-Anionic, Liquid)

Water	77.6
Sodium Tripolyphosphate	4.7
Trisodium Phosphate	4.7
Metso (anhydrous)	3.0
Monaterge 85	10.0

Procedure:

Add ingredients in the order listed with good agitation. pH (as is): 12.6.

Note:

Lower concentrations of **Monaterge 85** may be used with a proportional reduction in cleaning and rinsing properties.

No. 10

(Amphoteric)

Miranol® C2M-SF Conc.	22.0
Sodium Metasilicate Pentahydrate	30.0
Potassium Hydroxide (45%)	3.0
Dowanol PM	2.0
Water	43.0

No. 11

(Amphoteric)

Miranol® C2M-SF Conc.	22.00
Sodium Metasilicate Pentahydrate	30.00
Potassium Hydroxide (45% liquid)	3.00
Carbitol	1.66
Water	43.34

Automobile Solvent Glaze

(Nonionic)

A **GE SF96®** (1000)	3.0
Wax OM	1.4

Wax F	0.6
B Mineral Spirits	95.0

Procedure:
Heat the components of Part A to 100 C (212 F) until the waxes are completely melted. Use moderate agitation. Preheat Part B to 60 C (140 F) to prevent precipitation of the wax. Add Part B slowly to Part A with constant agitation.

Application:
Spread the film evenly with a clean, dry cloth. Allow the film to dry and then buff with a clean, dry cloth to a high gloss.

Warning:
When solvents are used, as described above, proper safety precautions must be observed. All solvents must be considered toxic and should be used only in well ventilated areas. Prolonged exposure to solvent vapors must be avoided. If flammable solvents are used, storage, mixing, and use must be in areas away from open flames or other sources of ignition. The selection of any solvent, particularly chlorinated hydrocarbon solvents, will require consideration of applicable OSHA, EPA, and other federal, state, and local regulations.

Automobile Vinyl Cleaner

Surfactant ET 0063	30.00
Ammonia	3.00
Butyl Cellosolve	5.00
Water (deionized)	62.00

Procedure:
Blend all components cold.

Note:
Aerosol products are usually applied neat to the surface. Nonaerosol products are used in accordance with the degree of soiling from neat to a dilution of 15:1.

Automobile Vinyl Top Cleaner

(Nonionic)

A	Veegum® HS	1.0
	CMC 7M	0.3
	Water	83.7
B	Butyl Cellosolve	5.0
	Tergitol NPX	8.0
C	Tetrapotassium Pyrophosphate	2.0

Procedure:

Add the **Veegum HS–CMC** dry blend to the water slowly, agitating continually until smooth. Add B and C in order to A mixing after each until smooth (avoid incorporation of air).

Packaging:

This formula is a low viscosity liquid ideally suited for pump spray dispensing.

Note:

Having excellent soil removal properties, this composition reflects the electrolyte and solvent stability of **Veegum HS**. When applied from a spray pump, the thixotropy imparted by the **Veegum HS** allows a fine spray and good foam without excessive running. When used along with a soft brush this formula efficiently cleans grease, heavy soil, or general dirt induced discolorations from the crevices and convolutions of textured vinyl tops.

Degreasing Agents

Formula No. 1

(Anionic, Metal)

Kerosene	80.0
Conco AAS Special #3	15.0
Water	5.0

No. 2

(Anionic, Spray Concentrate)

Aromatic 150	80.0
Monamulse 947	20.0

Procedure:
Add the ingredients in the order listed with moderate agitation.
Use concentration:
Aspirate at approximately 1 oz/gal.

No. 3

(Nonionic, Liquid)

Plurafac D-25	15
Plurafac A-24	5
Butyl Cellosolve	20
Kerosene	60

No. 4

(Nonionic-Anionic)

		Mixing Order
Steol CS-460	6	8
Ninol 1281	6	6
Tetrapotassium Pyrophosphate	4	5
Sodium Metasilicate	3	4
KOH	5	3
Stepanate X	5	2
Butyl Cellosolve	5	7
Water	66	1

Procedure:
Dissolve KOH, phosphate and silicate in water and **Stepanate X**. Mix in **Ninol 1281** and butyl Cellosolve. Add **Steol CS-460** and stir until uniform.

pH:	13.0
Gardner Color:	2+
Brookfield Visc.:	About 10 cps @ 25 C
Appearance:	Clear, light amber liquid

No. 5

(Amphoteric, Heavy Duty Spray)

Miranol® C2M-SF Conc.	3.3
Sodium Tripolyphosphate	1.4
Sodium Metasilicate Pentahydrate	2.7
Trisodium Phosphate	1.4
Tall Oil Fatty Acid	1.7
Potassium Hydroxide (45%)	1.0
Butyl Cellosolve	8.5
Water	80.0

pH: 12.8.

No. 6

(Gel, Brush-On)

Water	30.0
N Silicate	5.0
Monamulse 653-C	25.0
Heavy Aromatic Naphtha	40.0

This degreaser is sufficiently viscous so that it will cling to a brush for use on vertical surfaces. Because of the water content in the above formula, it is comparatively inexpensive. **Monamulse 653-C** permits the emulsified oily soil to be easily removed by spray rinsing, which normally follows a brush-on application. Also, application brushes can be rinsed clean under running water.

Butyl Cleaner and Degreaser

Formula No. 1

(Anionic)

Water	65.70
Trisodium Phosphate	4.35

Sodium Metasilicate (pentahydrate)	4.35
Potassium Hydroxide (45%)	3.50
ESI-Terge 320	3.60
ESI-Terge DDBSA	5.25
ESI-Terge SXS	4.50
Butyl Cellosolve	8.75

Procedure:

Add in order listed allowing 30 min of agitation after all ingredients are in.

pH:	13-13.5
Solids:	19-20
Active:	27-28
Visc.:	Water-like

No. 2

(Nonionic-Anionic)

Water	83.00
Trisodium Phosphate	3.00
Sodium Tripolyphosphate	3.00
ESI-Terge HA-20	5.00
*Potassium Hydroxide (45%)	1.00
Butyl **Cellosolve**	5.00

---

* Proper protective clothing should be used when handling potassium hydroxide.

Procedure:

Add in order listed with adequate agitation, allowing each material to dissolve before adding the **ESI-Terge HA-20** and butyl **Cellosolve**. Agitate until clear.

Solids:	6.5%
Active:	11.0%
pH:	12-13
Visc.:	10 cps LV #2 spindle @ 60 rpm @ 25 C

Solvent Cleaner

Formula No. 1

(Anionic)

Water	25.0
Isopropanol	25.0
Monawet MO65-150	10.0
Monamulse 947	40.0

Procedure:
Add the ingredients in the order listed with moderate agitation.

Use concentration:
Aspirate at approximately 1 oz/gal.

Note:
This is an excellent oil or grease remover with good rinsing properties. It has the additional advantage of forming fast breaking emulsions at use dilutions. For example, when used as an offshore oil rig cleaner the disposal wastes can be collected, and the oil easily separated before returning the water portion to the sea.

No. 2

(Anionic, Industrial)

Monamine ALX-100S	8.0
Chlorothene NU	5.0
Water	87.0

Note:
Thorough mixing will result in a coarse, but very stable emulsion cleaner for greasy machine parts which are not attacked by the chlorinated solvent.

No. 3

(Anionic, For Combustion Chambers)

Aromatic 100	70.0
Butyl Cellosolve	10.0
Monamulse 653-C	5.0

Monafax H-15	5.0
Diethanolamine	2.0
Water	8.0

Note:

The above formula is a thin, clear water-in-oil emulsion which blooms when added to water.

No. 4

(Anionic)

Perchloroethylene	30.0
Monafax H-15	2.5
Monamid 150-ADY	2.5
Water	65.0

No. 5

(Nonionic–Anionic, Water Dispersible)

Cropol 75	2.00
Volpo T5	1.25
Kerosene	96.50

Procedure:

Blend all components by stirring.

	No. 6	No. 7
	(Nonionic–Anionic, Water Dispersible)	
Kerosene	90.00	70.00
Solvent Naphtha	—	20.00
Volpo O5 and T5	7.00	7.00
Crillon LDE	3.00	3.00

No. 8

(Nonionic-Anionic, Spray)

Water	94.00
Monamine 779	3.00
Butyl Cellosolve	3.00

Procedure:
Mix ingredients in order listed.

No. 9

Varsol	90.0
Conco Emulsifier PO	10.0

Note:
This combination may be reduced with six parts of kerosene.
In addition, a 1% water emulsion shows good stability.

Dispersible Cleaners

(Cationic)

Water dispersible cleaners involving hydrocarbon solvents may also be based on cationic systems such as **Crodamet 1.T.2** and **Crodamet 1.S.5**. A combination of these two surfactants at a ratio of 1:2 and consisting of 5-10% of the total formula produces excellent dispersions. Further improvement in dispersing properties result from the inclusion of a suitable nonionic such as **Volpo T5**.

Crodafos phosphate esters may also be employed as dispersing or emulsifying agents. These materials also confer anticorrosion properties to the cleaner.

Cleaners based on chlorinated solvents may be rendered water dispersible by the addition of approximately 7% of **Crodafos T2, T5**, or **T10** acid. Surfactant **ET 0063** is also useful for emulsifying such solvents.

Jewelry Cleaners

Formula No. 1

Warm soap and water is applied with a toothbrush. The article is held flat on a table with one hand while the other hand taps the article with the toothbrush so that the bristles get into and around the design of the jewelry. This method is suitable for removing ordinary dust and dirt, although regular cleaning may be necessary to remove all the accumulated film.

No. 2

Kerosene	1 oz
Carbon Tetrachloride	3 oz
Citronella Oil	2 dr

Procedure:

Mix and apply to the parts to be cleaned with a soft cloth. Then polish.

Chapter VI

SPECIALTY CLEANERS

Bottle Cleaners

Formula No. 1

(Anionic)

Nacconol 90F	3
Caustic Soda	54
Trisodium Phosphate (anhydrous)	6
Sodium Tripolyphosphate (anhydrous)	6
Tetrasodium Ethylenediaminetetraacetate (dihydrate)	2
Water	1000

No. 2

(Anionic)

Nacconol 90F	3
Caustic Soda	50
Sodium Metasilicate (pentahydrate)	4
Trisodium Phosphate (monohydrate)	6
Sodium Tripolyphosphate (anhydrous)	6
Tetrasodium Ethylenediaminetetraacetate (dihydrate)	2
Water	1000

No. 3

(Anionic, Label Remover)

Nacconol 90F	0.1
Caustic Soda	6.0

Water	93.0
Disodium Ethylenediaminetetraacetate (dihydrate)	0.9

No. 4

(Amphoteric)

Miranol® JEM Conc.	0.01- 0.02
Sodium Hydroxide (50%)	6.00-12.00
Gluconic Acid	0.20- 0.60
Water	93.79-87.38

No. 5

(Nonionic-Amphoteric)

Sodium Hydroxide (50%)	12.000
Miranol® JEM Conc.	0.015
Igepal CO-730	0.005
Sodium Gluconate	0.200
Ethylenediamine Tetraacetic Acid	0.050
Water	87.730

No. 6

(Amphoteric)

Miranol® C2M-SF Conc.	1.0
Carbitol	1.0
Sodium Hydroxide Flakes	20.0
Water	78.0

Note:

This formulation is not suitable for use in high pressure bottle washing machines because of its foaming properties.

The surface tension of the above formulation is 37.7 dynes/cm, whereas the surface tension of a 20% sodium hydroxide solution alone is 74 dynes/cm. When the above formulation is diluted to contain 4% sodium hydroxide, the surface tension is 35.8 dynes/cm and when further diluted to contain 3% sodium hydroxide, the surface tension is 36.6 dynes/cm. By contrast the surface tension of 3 and 4% sodium hydroxide solutions containing no **Miranol® C2M-SF Conc.** is between 70 and 74 dynes/cm.

Leather Dyeing

(Cationic)

Arquad® 2C-75	1
Ethomeen® T/15	3
Mineral Oil	48
Neat's-foot Oil	48

Procedure:

The components are mixed to form a solution. This concentrate can be added to the drum containing wet leather.

Note:

The properties of leather as regards coloring with dyestuffs can be altered by pretreatment with an aqueous solution of **Arquad 12, Arquad 16,** or **Arquad S.** When the leather is drummed with the **Arquad** solution, it often shows an increased affinity for the dyestuff.

Dye Substantivity

Small amounts of **Accoquat 2C75** when added to dye baths produces substantivity of dyes for many textile materials.

Leather Stain Remover

A solution for removing stains from the flesh side of leather is composed of the following:

Water	250 cc
Oxalic Acid	3 g

Procedure:

Mix the ingredients together.

Degreaser for Hides

Formula No. 1

(Nonionic)

Water	90
Benzene	5
Polyglycol 200 Monooleate	5

Procedure:
Mix all ingredients together until uniform.

No. 2

(Nonionic)

Water	87
Benzene	5
Polyglycol 200 Monostearate	2
Olive Oil	4

Procedure:
Mix all ingredients together until uniform.

Teat Dip Concentrate

(Nonionic)

Iodofors (**Antarox VRO 20**)	8.50
Solan E	4.00
Croderol G 7000	10.00
Phosphoric Acid (85%)	1.50
Water (deionized)	76.00

This preparation is diluted 1:2 with water before application to the teat or on udder cloths as a solution of approximately ½ fluid ounce to one gallon of water. This is recommended to keep cloths disinfected during milking.

Teat and Udder Cleaner

(Nonionic)

Cetrimide BP	0.50
Croderol G 7000	10.00
Solan E	1.00
Water (deionized)	88.50

Milk Salve

Formula No. 1

(Nonionic, Cream)

Ceto Stearyl Alcohol	10.00
Polawax	4.00
Yeoman Lanolin	1.00
Liquid Paraffin (BP)	5.00
Preservative	q.s.
Cetrimide BP	0.50
Croderol G 7000	1.00
Water (deionized)	78.50

Procedure:
Heat water phase and oil phase separately to 70 C and combine with agitation.

No. 2

(Nonionic, Lotion)

Cosmowax	3.00
Liquid Paraffin (light)	8.00
Hartolan	0.20
Preservative	q.s.
Croderol G 7000	2.00
Cetrimide BP	0.50
Water (deionized)	86.30

Procedure:
Heat oil phase and water phase separately to 70 C and combine with agitation.

Pet Shampoo

Formula No. 1

(Nonionic-Anionic)

A **Veegum**	1.0
Water	42.9
Citric Acid	0.4
Igepon AC-78 (83% solids)	18.0
Igepon TC-42 (24% solids)	25.0
B Cetyl Alcohol	1.8
Glyceryl Monostearate A.S.	5.9
Solulan 98	3.5
C Synergized Pyrethrins (50% piperonyl butoxide and	
10% pyrethrins)	0.5
D **Vancide 89RE**	1.0

Procedure:

Add the **Veegum** to the water slowly, agitating continually until smooth. Add rest of A and heat to 75 C. Mix B and heat to 80 C. Add B to A, mixing until cool. Add C to A and B and mix. Add D to a small portion of the cream, disperse thoroughly. Add this concentrate to the remainder of the shampoo. Mix until uniform.

Note:

The final pH should be about 5.0.

No. 2

(Amphoteric)

Miranol® C2M Conc.	25.0
Lauric Diethanolamide (high active)	4.0
Hexylene Glycol	2.0
Pyrocide Intermediate 5192[1]	0.5
Water	68.5

---

[1] McLaughlin Gormley King Co.

Note:

Adjust pH to 7.0-8.0 with hydrochloric acid.

No. 3

(Amphoteric)

Miranol® C2M Conc.	25.0
Hexylene Glycol	2.0
Pyrocide Intermediate 5192	0.5
Water	72.5

Dust Control Formulation

Formula No. 1

(Nonionic-Cationic)

Accoquat 2C75	2.5
Accosperse S-35	2.5
Mineral Oil	95

Accoquat 2C75 can be blended with mineral oil and a soya amine ethoxylate to give a cationic soluble mineral oil concentrate. Diluted with water this formulation gives an excellent dust control product.

No. 2

(Cationic)

Monazoline T	30
Tergitol 15-S-7	20
Mineral Oil (light)	50

This substantive oil concentrate will improve the pick-up properties of dusting mops when applied as a 2% aqueous solution.

No. 3

(Nonionic-Cationic, Oil)

Mineral Oil	96
Variquat K300	2
Varionic L202	2

Procedure:

Mix all ingredients until uniform.

The composition exhibits a degree of germicidal and fungicidal activity because of the biocidal properties of the **Variquat K300.**

Antistatic Agent

Formula No. 1

(Nonionic–Cationic)

Isopropanol	5-10
Alkazine O	1- 2
Acetic Acid	0.2- 0.4
Alkamox LO	0.1
Color and Preservative	q.s.
Water	to 100

No. 2

(For Carpet)

Q-18-2 (melted)	3
Isopropanol	7
Water (warm)	90

One gallon will treat 1000 sq. ft of carpet.

Note:

Q-18-2 minimizes soil pick-up while providing good antistatic properties.

Q-14-2 is a more effective antistatic agent but will leave a liquid film on the fiber that gives more dirt pick-up.

No. 3

(For Textile)

Q-18-2, Q-C-2, Q-14-2, and **PA-14** salts are used during textile processing at 0.2% levels to reduce static on wool and most synthetic fibers. Combination antistatic-lubricants are widely used. Many cationic emulsions are used to exhaust a hydrophobe, such as polyethylene waxes, on the textile and to achieve antistatic properties at the same time.

No. 4

(For Plastic)

Internal antistatic agents such as **E-14-2** and **E-17-2**, when incorporated in polyethylene resins, migrate to the surface to provide antistatic properties and reduce dust pick-up on molded products like bleach bottles.

Q-17-2, Q-14-2, and **PA-14** Acetate are used as external antistatic agents for olefin, acrylic, and styrene plastic products such as phonograph records, plastic films, bottles, and signs.

No. 5

(For Electrostatic Painting)

Electrostatic spray painting requires a conductive surface and a conductive paint. Paint is adjusted to the proper conductivity by blending polar and nonpolar solvents plus compatible conductivity additives, such as Tomah's **Q-C-2, Q-14-2, Q-S,** and **Q-17-2,** used at ¼–1% levels.

For electrostatic spraying of nonconductive surfaces like wood or plastic, a pretreatment of **Q-D-T** or **Q-14-2** will make the surface conductive.

Acid Corrosion Inhibitor

Formula No. 1

(Hydrochloric Acid)

E-S-15	60
Propargyl Alcohol or Butynediol	10
Isopropanol	30

or

PA-17	83
Propargyl Alcohol	17

To achieve 98-99% protection, use 1% Inhibitor based on 100% active acid. For steel mill pickling, modifications are required to achieve proper pickling speed, solubility, and foaming. **Q-14-2** shows promise in steel mill pickling inhibitors.

No. 2

(Sulfuric Acid)

Q-18-15	60
Dibutyl Thiourea	20
Isopropanol	20

Note:

0.2% inhibitor in 10% H_2SO_4 at 25 C will give 99% protection to 1008 steel.

No. 3

(Phosphoric, Citric, Sulfamic Acids)

Q-14-2	60
Dibutyl Thiourea	20
Isopropanol	20

Fingerprint Remover Base

Alox 904	9.0
100 SUS @ 100 F Coastal Oil	35.0
Kerosene	48.5
Water	7.0
Butyl **Cellosolve**	0.5

Note:

1. *Oils Selected*

The clarity of the fingerprint remover depends on the oil selected to some extent. The exact limitations to be placed on viscosity, source, or refining method cannot be specified in detail. Both acid-treated and solvent-refined oils have been used successfully. Viscosity of the oil does not appear to be a critical factor in stability or performance.

2. *Kerosene Selected*

When a high flash point (200-250) fingerprint remover compound is required, the kerosene can be replaced in the formula by a suitable process or heavy (#3) fuel oil. Process oils or fuel oils having some aromatic content appear to be most desirable.

3. *Butyl Cellosolve*

Since oils from different sources behave differently, the exact amount of butyl **Cellosolve** used depends on the oil and/or kerosene selected. Clarity of solution could be adjusted with butyl **Cellosolve** for all the oils evaluated. Using the proper oil and kerosene, the recommended formulation required only 0.5% butyl Cellosolve.

Procedure:

Alox 904 is added to the cold oil (no heating necessary) with good stirring until it is completely and uniformly dissolved. Agitation is continued during the further addition of kerosene and water. Stirring is continued for 15-30 min longer, and, as the final step, add enough butyl **Cellosolve** to produce a clear mixture. As described previously, the amount of butyl **Cellosolve** required may have to be determined for the oil/kerosene blend selected. Usually 0.5% is adequate.

While the above procedure can be followed without employing heat, the oil can be warmed, if desired, to hasten solution of **Alox 904**. However, if heated, the oil/904 blend must be cooled before addition of the butyl **Cellosolve**.

Nicotine Finger-Stain Remover

Hydrochloric Acid (pure)	0.4
Glycerin	10.0
Rose Water (triple)	90.0

Procedure:

Combine ingredients with agitation until homogeneous.

To use the lotion, saturate a piece of cotton with it and rub gently over the stained area. Several applications may be necessary before desired results are obtained.

Soot Remover

Formula No. 1

Common Salt	85.0
Copper Sulfate	8.0
Zinc Dst	7.0

No. 2

Coarse Rock Salt	125.0
Zinc Dust	2.0
Copper Sulfate	2.0

No. 3

Zinc Dust	50.0
Sodium Chloride	45.0
Sawdust	5.0

Procedure:
 Mix the ingredients of each formula until smooth.
 Throw approximately 5 oz of soot remover onto fire.

Aerosol Foam Cleaner for Plastic Upholstery

(Anionic)

		% in Aerosol
Concentrate:		90.0
A Dow Corning 35B Emulsion	5.55	
Sarkosy I NL-30	1.67	
Duponol ME	2.22	
Sipon LT-6	1.67	
Trisodium Phosphate	3.34	
Water	58.45	
B Stoddard Solvent	11.00	
Oleic Acid	1.90	

C Triethanolamine	0.60	
Water	13.60	
Propellant:		10.0
Isotron 12	50.0	
Isotron 114	50.0	

Procedure:
Blend Parts B and C separately and then together in a high-shear mixer. (Warning: Flammable.) Blend A separately, then add B and C to A.

Note:
This cleaner is suitable for vinyl, Saran, and other nonabsorptive plastic.

Package:
Pressure fill in lacquer-lined cans with foam valves.

Directions for Use:
Shake well. Foam a small amount on surface to be cleaned. Rub with a soft cloth to remover dirt and to produce a sheen.

Warning:
Contents under pressure. Do not puncture. Exposure to heat or prolonged exposure to sun may cause bursting. Do not throw into fire or incinerator. Keep from children.

Rug Spot Remover

(Anionic)

		Mixing Order
Ninate 415	3.0	2
Butyl Cellosolve	4.0	3
Carbitol Solvent	4.0	4
Water	89.0	1

Procedure:
Mix **Ninate 415** with water; add remaining ingredients.

pH:	4.0–5.0
Brookfield Visc.:	About 5 cps @ 25 C
Appearance:	Opaque liquid

Spot and Stain Remover

Formula No. 1

(Nonionic)

		% in Aerosol
Concentrate:		85.0
Renex 20	70.0	
"N" Brand Silicate	0.5	
Water	29.5	
Propellant:		15.0
Isotron 12	40.0	
Isotron 114	60.0	

Procedure:

Mix ingredients together until homogeneous. Product should be pressure filled.

Note:

A foam-type cleaner for both water soluble and grease stains.

Package:

Lacquer-lined container with a foam-type valve.

Directions for use:

Place fabric on an absorbent material to remove excess water and to avoid formation of ring. Apply directly to the spot to be cleaned. Use a soft brush and work foam into the fabric.

Warning:

Contents under pressure. Do not puncture. Exposure to heat or prolonged exposure to sun may cause bursting. Do not throw into fire or incinerator. Keep from children.

No. 2

(Nonionic)

Tergitol N.P.X.	0.1 lb
Carbitol	25.6 cc
Isopropyl Alcohol	19.2 cc
Cleaner's Naphtha	102.4 cc

Procedure:
Mix **Carbitol**, isopropyl alcohol and naphtha. Add **Tergitol** and stir until a clear solution is obtained.

Warning:
Avoid open flames, when mixing or using this formula.

Note:
This product can remover lipstick stain from fabrics.

No. 3

(Powder)

Aerothene TT	45.00
Dow-Per	15.00
Isopropyl Alcohol	9.95
Perfume	0.05
Microcel C	5.00
Propellant 12	25.00

Suggested Valve: Risdon #5832 0.020″ stem/0.138″ body
Suggested Actuator: Risdon #5832 Actuator (EH-16) 0.030″
Suggested Perfumes: Fleuroma Bouquet #686, Roure du Pont Bouquet #A5147.

Procedure:
First mix the solvents and perfume. Then slowly mix in the powder. Mix thoroughly to get an even dispersion. Mix **Aerothene TT** and perfume then add the other ingredients. Stir until uniform. Fill in aerosol can or bottle and charge with the Propellant 12.

To apply:
Place an absorbent cloth on the underside of the spot. Apply a liberal coating to the topside of the cloth. Allow the powder to dry thoroughly before brushing off.

Rust Inhibitor

Formula No. 1

(Anionic)

SAE 40 Motor Oil	25.0

Monawet MO65-150	0.5
Monacor 39	3.0
Aromatic 150	71.5

Note:

Monawet MO65-150's good wetting properties assist in penetrating and freeing frozen bolts, etc. It also acts as a water displacing agent. **Monacor 39** provides additional corrosion inhibiting properties.

No. 2

(Anionic)

Mazon R.I. #325	25
TEA	25
Maphos 8135	25
Water	25

No. 3

(Anionic)

Mazon R.I. #325	20
TEA	20
Maphos 8135	20
Water	10
Hexylene Glycol	10
Mazon 86 LF	20

No. 4

(Nonionic-Anionic, Water-Soluble)

Mazon R.I. #325	15
TEA	15
Maphos 8135	10
Macol HB 3500	10
Water	50

Rust Penetrant

		% in Aerosol
Concentrate:		50.0
Acme Oil B	70.0	
Reprol Seal Oil	30.0	
Propellant:		50.0
Isotron 12	100.0	

Procedure:
May be pressure or cold filled. This formulation loosens rusted and frozen metal parts.

Package:
Tinplate container with paint-type valve.

Directions for use:
Spray affected area and allow to stand 1-2 min before attempting to loosen.

Warning:
Contents under pressure. Do not puncture. Exposure to heat or prolonged exposure to sun may cause bursting. Do not throw into fire or incinerator. Keep from children.
Flammable mixture. Harmful if swallowed.

Rust Remover

Formula No. 1

(Amphoteric, Gel)

Miranol® JS Conc.	4.0
Caustic Soda (50% liquid)	50.0
Sodium Gluconate	6.0
Versene® FE-3 Specific	1.5
Water	38.5

No. 2

(Amphoteric, Liquid)

Miranol® JEM Conc.	2.0
Carbitol	2.0

Potassium Hydroxide (45% liquid)	78.0
Triethanolamine	12.0
Water	6.0

Aerosol Dewatering Rust Preventive Spray

Pentalan	5.00
Pool 3 Mineral Oil	10.00
White Spirit	80.50
Butyl Cellosolve	3.00
Cropol 75	1.06
Petroleum Sulfonate M50	0.50

Pack: Concentrate 40%
          Propellant 12/11 (50:50) 60%

Nonacidic Scale and Slime Remover

(Anionic)

Water	78.77
Sodium Hexametaphosphate	8.50
Versene 100	10.60
ESI-Terge 320	2.13

Procedure:
Dissolve sodium hexametaphosphate in water. After solution clears and is free of salt, add **Versene 100** and **ESI-Terge 320**.

| pH: | 7.5-8.5 |
| Activity: | 14.5-15.5% |

Heat-Formed Oxides or Scale Remover

Formula No. 1

(Liquid Product)

Caustic Soda Liquor (47.5% wt)	46.00
Sodium Heptonate (industrial)	26.00
Sodium Cyanide Liquor (30% wt)	28.00

Concentration in use: 500/800 g/l

Temperature:
55 C maxiumum. Above this temperature decomposition of cyanide takes place with liberation of ammonia.

No. 2

(Powder Product)

Caustic Soda (flake)	56.00
Sodium Heptonate Dihydrate	22.00
Sodium Cyanide Granula	22.00

Concentration in use: 200/300 g/l

Temperature:
55 C maximum. Above this temperature decomposition of cyanide takes place with liberation of ammonia.

Note:
As an alternative to the use of cyanides (which may be objectionable) it is possible to use amino carboxylate-type sequestering agents such as **Crodaquest EDTA** and **DTPA**. Triethanolamine is also an effective additive and can be incorporated at a level of about 10% of the caustic soda present, either as the free base or in the form of triethanolamine phosphate or hydrochloride.

No. 3

(Liquid Product)

Caustic Soda Liquor (47.5% wt)	43.00
Sodium Heptonate (industrial)	32.00
Crodaquest EDTA	25.00

No. 4

(Powder Product)

Caustic Soda (flake)	50.00
Sodium Heptonate Dihydrate	25.00
EDTA Di Sodium Salt	25.00

Concentration in use: 300 g/l

Temperature: 95 C

Chain Lubricant

Formula No. 1

(Amphoteric)

Miranol® JEM Conc.	20.0
Ethylenediamine Tetra-Acetic Acid	5.0
Sodium Hydroxide (50% liquid)	2.0
Water	73.0

No. 2

(Amphoteric)

Miranol® JEM Conc.	20.0
Water	80.0

No. 3

(Amphoteric)

Caprylic Acid } neutralized to pH 11.6	10.0
Oleic Acid	10.0
Miranol® JEM Conc.	10.0
Water	70.0

Aerosol Deicer

Cropol 60	0.20
Isopropyl Alcohol	54.50
Ethylene Glycol	20.00
Propellant 12	25.00

Aerosol Battery Terminal Protective/Leak Indicator

White Spirit	50.00
Petroleum Jelly	40.00
Methyl Orange (indicator)	6.00
Crill 4	4.00

Procedure:

The concentrate produced above needs slight warming in order to produce a reasonably homogeneous system. It is aerosol packed in a ratio of 70 parts concentrate to 30 parts Propellant 12.

Steam Cleaning Compound

Formula No. 1

(Anionic, Medium-Duty Steam)

Grade 40 Silicate of Soda	38
Potassium Hydroxide (45%)	32
Potassium Pyrophosphate (60%)	29
Monawet SNO-35	1

Use level — 1 oz/gal of water.

No. 2

(Anionic, For Carpet)

Water	78.50
Sodium Tripolyphosphate	4.00
KOH (45%)	7.00
ESI-Terge 330	10.00
ESI-Terge T-60	0.50

Procedure:

Add ingredients in order listed.

pH:	8.0–8.5
Active:	17.45%
Visc.:	Water like

Warning:
Precaution should be taken in handling KOH 45%.

No. 3

(Anionic)

Water	93.00
Potassium Hydroxide (90%)	5.00
ESI-Terge HA-20	1.00
Versene 100	1.00

Procedure:
Add in order listed.

Specifications:
Solids:	5.9
Active:	5.9
pH:	13.25–13.75
Visc.:	100 cps at room temperature–LV #1 spindle @ 60 rpm @ 25 C

Note:
Solids may be increased proportionally or sodium metasilicate may be added. For 5 parts combination of alkali or silicates use 1 part of **ESI-Terge.**

Warning:
Proper protective clothing should be used when handling potassium hydroxide.

No. 4

(Anionic)

Water	93.00
Sodium Hydroxide or Potassium Hydroxide (97%)	5.00
ESI-Terge 330	1.00
Versene 100	1.00

Procedure:
Add in order listed.

Specifications:
Solids:	6.5-7.0
Active:	6.5-7.0
pH:	13.0-13.5
Visc.:	Water like

Note:
Solids may be increased proportionally or sodium metasilicate may be added. For 5 parts combination of alkali or silicates use 1 part of **ESI-Terge.**

Warning:
Protective clothing should be used when handling potassium hydroxide. Heat is given off while reacting.

No. 5

(Nonionic, For Rugs)

Alkasurf LA-16	18-22
Alkasurf NP-11	5- 8
Alkasurf TDA-7.5	5- 8
Alkamide 2112	2- 4
Pine Oil	3- 5
Clean Sol 34	10-20
Color and Perfume	q.s.
Water	to 100

No. 6

(Nonionic, Powdered)

Soda Ash	65
Triton X-100	2
Alkali Surfactant	2
Sodium Metasilicate (anhydrous)	15
NTA	2
Sodium Gluconate	4
Flake Caustic Soda (NaOH)	10

Chelating powder: adequate for 6 oz/gal dilution in 400 ppm hard water.
pH at dilution: 13.6 at 20%, 13.3 at 10%, and 13.1 at 5% concentration.

No. 7

(Nonionic-Anionic)

		Mixing Order
Makon 10	3-5	4
Sodium Silicate	10	3
Stepanate X	3-5	2
Water	Balance	1

Procedure:
Dissolve the silicate in water and **Stepanate X**. Add the **Makon 10**. Stir until uniform.

pH:	12.0
Brookfield Visc.:	About 5 cps @ 25 C
Appearance:	Colorless, slightly hazy liquid

No. 8

(Amphoteric, Heavy-Duty Liquid)

Potassium Hydroxide (45%)	55.0
Kasil #1	32.0
Gluconic Acid (50%)	4.0
Phosphoric Acid (75%)	8.0
Miranol® J2M Conc.	1.0

Procedure:
Mix in the order listed.

Warning:
Phosphoric acid must be added very slowly with constant stirring to avoid spattering. The solution is clear and slightly yellow.

No. 9

(Amphoteric, Concentrated Liquid)

Monateric Cy Na (50%)	15.0
Sodium Metasilicate · $5H_2O$	20.0
Potassium Hydroxide (45%)	22.0
Water	43.0

No. 10

(Amphoteric, Medium-Duty Liquid)

Water	37.0
Potassium Hydroxide (45%)	20.0
Kasil #1	22.0
Potassium Carbonate	12.0
Gluconic Acid	8.0
Miranol® J2M Conc.	1.0

Procedure:
Mix in the order listed.

	No. 11	No. 12
	(Amphoteric)	
Antaron FC-34	20	30
Sodium Metasilicate	10-20	5
Sodium Chloride	—	3
Water	70-60	62

Note:
Formula 11 is a general heavy-duty cleaner; Formula 12 makes a good aluminum cleaner. Aluminum parts may be cleaned by immersion in this compound.

No. 13

(Amphoteric)

Miranol® C2M-SF Conc.	15.0
Starso	62.0

Potassium Hydroxide (45% liquid)	10.0
Water	13.0

No. 14

(Amphoteric)

Miranol® C2M-SF Conc.	10.0
Starso	90.0

No. 15

(Amphoteric)

Miranol® C2M-SF Conc.	15.0
Sodium Metasilicate Pentahydrate	20.0
Water	65.0

No. 16

(Amphoteric)

Miranol® C2M-SF Conc.	15.0
Sodium Metasilicate Pentahydrate	20.0
Potassium Hydroxide (45% liquid)	11.0
Water	54.0

No. 17

(Amphoteric)

Miranol® C2M-SF Conc.	15.0
Sodium Metasilicate Pentahydrate	20.0
Potassium Hydroxide (45% liquid)	22.0
Water	43.0

No. 18

(Amphoteric)

Miranol® C2M-SF Conc.	15.0
Sodium Metasilicate Pentahydrate	40.0
Water	45.0

No. 19

(Amphoteric)

Miranol® C2M-SF Conc.	15.0
Sodium Metasilicate Pentahydrate	30.0
Water	55.0

No. 20

(Amphoteric)

Miranol® C2M-SF Conc.	15.0
Sodium Metasilicate Pentahydrate	30.0
Potassium Hydroxide (45% liquid)	11.0
Water	44.0

No. 21

(Amphoteric)

Miranol® C2M-SF Conc.	15.0
Sodium Metasilicate Pentahydrate	35.0
Water	50.0

No. 22

(Amphoteric)

Miranol® C2M-SF Conc.	15.0
Sodium Metasilicate Pentahydrate	50.0
Water	35.0

No. 23

Condensate XB Special	7.0
Sodium Tripolyphosphate	6.0
Sodium Metasilicate Pentahydrate	4.0
Water	83.0

No. 24

(Liquid Concentrate)

Water	65
NTA	1
Sodium Gluconate	4
KOH (45%)	25
Alkali Surfactant	3
Triton X-100	2

Procedure:

Combine the ingredients in the order given.

Wt/gal:		9.5 lb
Stability:	stable from 35 F to 120 F, usable to 212 F	
Chelating power:	adequate for 50 to 1 dilution in 400 ppm hard water	
pH at dilution:	as is	14.0
	5 to 1	13.3
	10 to 1	13.1
	20 to 1	12.7
	50 to 1	12.2

Dry Cleaner Charge Soap

	Formula No. 1	No. 2	No. 3	No. 4
		(Nonionic-Anionic)		
Alkasurf I.P.A.M.	40	40	40	40
Alkaphos-3	–	15	10	–
Alkamide CDO	–	–	10	10
Alkasurf LA-3	20	15	–	10
Stoddard Solvent	40	30	40	40

Note:

The water solubilizing capacity of **Alkasurf I.P.A.M.** permits its use as a base material for the preparation of dry cleaner charge soaps.

APPENDIX

pH Values

Acids	pH Value	Bases	pH Value
Hydrochloric Acid	1.0	Sodium Bicarbonate	8.4
Sulfuric Acid	1.2	Borax	9.2
Phosphoric Acid	1.5	Ammonia	11.1
Sulfuric Acid	1.5	Sodium Carbonate	11.6
Acetic Acid	2.9	Trisodium Phosphate	12.0
Alum	3.2	Sodium Metasilicate	12.2
Carbonic Acid	3.8	Lime, Saturated	12.3
Boric Acid	5.2	Sodium Hydroxide	13.0

pH Ranges of Common Indicators

	Useful pH Range
Thymol Blue	1.2 − 2.8
Bromphenol Green	2.8 − 4.6
Methyl Orange	3.1 − 4.4
Bromcresol Green	4.0 − 5.6
Methyl Red	4.4 − 6.0
Propyl Red	4.8 − 6.4
Bromcresol Purple	5.2 − 6.8
Brom Thymol Blue	6.0 − 7.6
Phenol Red	6.8 − 8.4
Litmus	7.2 − 8.8
Cresol Red	7.2 − 8.8
Cresolphthalein	8.2 − 9.8
Phenolphthalein	8.6 − 10.2
Nitro Yellow	10.0 − 11.6
Alizarin Yellow R	10.1 − 12.1
Sulfo Orange	11.2 − 12.6

International Atomic Weights

Element	Symbol	Atomic Number	Atomic Weight
Actinium	Ac	89	227
Aluminum	Al	13	26.98
Americium	Am	95	243
Antimony	Sb	51	121.76
Argon	A	18	39.944
Arsenic	As	33	74.91
Astatine	At	85	210
Barium	Ba	56	137.36
Berkelium	Bk	97	249
Beryllium	Be	4	9.013
Bismuth	Bi	83	209.00
Boron	B	5	10.82
Bromine	Br	35	79.916
Cadmium	Cd	48	112.41
Calcium	Ca	20	40.08
Californium	Cf	98	249.00
Carbon	C	6	12.011
Cerium	Ce	58	140.12
Cesium	Cs	55	132.91
Neodymium	ND	60	144.27
Neptunium	Np	93	237.00
Neon	Ne	10	20.183
Nickel	Ni	28	58.71
Niobium	Nb	41	92.91
Nitrogen	N	7	14.008
Osmium	Os	76	190.20
Oxygen	O	8	16
Palladium	Pd	46	106.4
Phosphorus	P	15	30.975
Platinum	Pt	78	195.09
Plutonium	Pu	94	242.00
Potassium	K	19	39.100
Praseodymium	Pr	59	140.92
Promethium	Pm	61	145
Protactinium	Pa	91	231
Radium	Ra	88	226.05
Radon	Rn	86	222
Rhenium	Re	75	186.22

International Atomic Weights (cont.)

Name	Symbol	Atomic Number	Atomic Weight	Name	Symbol	Atomic Number	Atomic Weight
Chlorine	Cl	17	35.457	Rhodium	Rh	45	102.91
Chromium	Cr	24	52.01	Rubidium	Rb	37	85.48
Cobalt	Co	27	58.94	Ruthenium	Ru	44	101.10
Copper	Cu	29	63.54	Samarium	Sm	62	150.35
Curium	Cm	96	245	Scandium	Sc	21	44.96
Dysprosium	Dy	66	162.51	Selenium	Se	34	78.96
Erbium	Er	68	167.27	Silicon	Si	14	28.09
Europium	Eu	63	152	Silver	Ag	47	107.880
Fluorine	F	9	19	Sodium	Na	11	22.991
Francium	Fr	87	223	Strontium	Sr	38	87.63
Gadolinium	Gd	64	157.26	Sulfur	S	16	32.066
Gallium	Ga	31	69.72	Tantalum	Ta	73	180.95
Germanium	Ge	32	72.60	Technetium	Tc	43	99
Gold	Au	79	197.00	Tellurium	Te	52	127.61
Hafnium	Hf	72	178.50	Terbium	Tb	65	158.93
Helium	He	2	4.003	Thallium	Tl	81	204.39
Holmium	Ho	67	164.94	Thorium	Th	90	232.05
Hydrogen	H	1	1.0080	Thulium	Tm	69	168.94
Indium	In	49	114.82	Tin	Sn	50	118.70

International Atomic Weights (cont.)

	Symbol	Atomic Number	Atomic Weight
Iodine	I	53	126.91
Iridium	Ir	77	192.20
Iron	Fe	26	55.85
Krypton	Kr	36	83.80
Lanthanum	La	57	138.92
Lead	Pb	82	207.21
Lithium	Li	3	6.940
Lutetium	Lu	71	174.99
Magnesium	Mg	12	24.32
Manganese	Mn	25	54.94
Mendelevium	Mv	101	256.00
Mercury	Hg	80	200.61
Molybdenum	Mo	42	95.95

	Symbol	Atomic Number	Atomic Weight
Titanium	Ti	22	47.90
Tungsten	W	74	183.86
Uranium	U	92	238.07
Vanadium	V	23	50.95
Xenon	Xe	54	131.30
Ytterbium	Yb	70	173.04
Yttrium	Y	39	88.92
Zinc	Zn	30	65.38
Zirconium	Zr	40	91.22

Temperature Conversion Tables

F	C	F	C	F	C	F	C	F	C
-40	-40.0	9	-12.8	58	14.4	107	41.7	156	68.9
-39	-39.4	10	-12.2	59	15.0	108	42.2	157	69.4
-38	-38.9	11	-11.7	60	15.6	109	42.8	158	70.0
-37	-38.3	12	-11.1	61	16.1	110	43.3	159	70.6
-36	-37.8	13	-10.6	62	16.7	111	43.9	160	71.1
-35	-37.2	14	-10.0	63	17.2	112	44.4	161	71.7
-34	-36.7	15	-9.4	64	17.8	113	45.0	162	72.2
-33	-36.1	16	-8.9	65	18.3	114	45.6	163	72.8
-32	-35.6	17	-8.3	66	18.9	115	46.1	164	73.3
-31	-35.0	18	-7.8	67	19.4	116	46.7	165	73.9
-30	-34.4	19	-7.2	68	20.0	117	47.2	166	74.4
-29	-33.9	20	-6.7	69	20.6	118	47.8	167	75.0
-28	-33.3	21	-6.1	70	21.1	119	48.3	168	75.6
-27	-32.8	22	-5.6	71	21.7	120	48.9	169	76.1
-26	-32.2	23	-5.0	72	22.2	121	49.4	170	76.7
-25	-31.7	24	-4.4	73	22.8	122	50.0	171	77.2
-24	-31.1	25	-3.9	74	23.3	123	50.6	172	77.8
-23	-30.6	26	-3.3	75	23.9	124	51.1	173	78.3
-22	-30.0	27	-2.8	76	24.4	125	51.7	174	78.9
-21	-29.4	28	-2.2	77	25.0	126	52.2	175	79.4
-20	-28.9	29	-1.7	78	25.6	127	52.8	176	80.0
-19	-28.3	30	-1.1	79	26.1	128	53.3	177	80.6
-18	-27.8	31	-0.6	80	26.7	129	53.9	178	81.1
-17	-27.2	32	0.0	81	27.2	130	54.4	179	81.7
-16	-26.7	33	0.6	82	27.8	131	55.0	180	82.2
-15	-26.1	34	1.1	83	28.3	132	55.6	181	82.8
-14	-25.6	35	1.7	84	28.9	133	56.1	182	83.3
-13	-25.0	36	2.2	85	29.4	134	56.7	183	83.9
-12	-24.4	37	2.8	86	30.0	135	57.2	184	84.4
-11	-23.9	38	3.3	87	30.6	136	57.8	185	85.0
-10	-23.3	39	3.9	88	31.1	137	58.3	186	85.6
-9	-22.8	40	4.4	89	31.7	138	58.9	187	86.1
-8	-22.2	41	5.0	90	32.2	139	59.4	188	86.7
-7	-21.7	42	5.6	91	32.8	140	60.0	189	87.2
-6	-21.1	43	6.1	92	33.3	141	60.6	190	87.8

Temperature Conversion Tables (cont.)

F	C	F	C	F	C	F	C	F	C
-5	-20.6	44	6.7	93	33.9	142	61.1	191	88.3
-4	-20.0	45	7.2	94	34.4	143	61.7	192	88.9
-3	-19.4	46	7.8	95	35.0	144	62.2	193	89.4
-2	-18.9	47	8.3	96	35.6	145	62.8	194	90.0
-1	-18.3	48	8.9	97	36.1	146	63.3	195	90.6
0	-17.8	49	9.4	98	36.7	147	63.9	196	91.1
1	-17.2	50	10.0	99	37.2	148	64.4	197	91.7
2	-16.7	51	10.6	100	37.8	149	65.0	198	92.2
3	-16.1	52	11.1	101	38.3	150	65.6	199	92.8
4	-15.6	53	11.7	102	38.9	151	66.1	200	93.3
5	-15.0	54	12.2	103	39.4	152	66.7	201	93.9
6	-14.4	55	12.8	104	40.0	153	67.2	202	94.4
7	-13.9	56	13.3	105	40.6	154	67.8	203	95.0
8	-13.3	57	13.9	106	41.1	155	68.3	204	95.6
205	96.1	254	123.3	303	150.6	352	177.8	401	205.0
206	96.7	255	123.9	304	151.1	353	178.3	402	205.6
207	97.2	256	124.4	305	151.7	354	178.9	403	206.1
208	97.8	257	125.0	306	152.2	355	179.4	404	206.7
209	98.3	258	125.6	307	152.8	356	180.0	405	207.2
210	98.9	259	126.1	308	153.3	357	180.6	406	207.8
211	99.4	260	126.7	309	153.9	358	181.1	407	208.3
212	100.0	261	127.2	310	154.4	359	181.7	408	208.9
213	100.6	262	127.8	311	155.0	360	182.2	409	209.4
214	101.1	263	128.3	312	155.6	361	182.8	410	210.0
215	101.7	264	128.9	313	156.1	362	183.3	411	210.6
216	102.2	265	129.4	314	156.7	363	183.9	412	211.1
217	102.8	266	130.0	315	157.2	364	184.4	413	211.7
218	103.3	267	130.6	316	157.8	365	185.0	414	212.2
219	103.9	268	131.1	317	158.3	366	185.6	415	212.8
220	104.4	269	131.7	318	158.9	367	186.1	416	213.3
221	105.0	270	132.2	319	159.4	368	186.7	417	213.9
222	105.6	271	132.8	320	160.0	369	187.2	418	214.4
223	106.1	272	133.3	321	160.6	370	187.8	419	215.0
224	106.7	273	133.9	322	161.1	371	188.3	420	215.6
225	107.2	274	134.4	323	161.7	372	188.9	421	216.1

Temperature Conversion Tables (cont.)

F	C	F	C	F	C	F	C	F	C
226	107.8	275	135.0	324	162.2	373	189.4	422	216.7
227	108.3	276	135.6	325	162.8	374	190.0	423	217.2
228	108.9	277	136.1	326	163.3	375	190.6	424	217.8
229	109.4	278	136.7	327	163.9	376	191.1	425	218.3
230	110.0	279	137.2	328	164.4	377	191.7	426	218.9
231	110.6	280	137.8	329	165.0	378	192.2	427	219.4
232	111.1	281	138.3	330	165.6	379	192.8	428	220.0
233	111.7	282	138.9	331	166.1	380	193.3	429	220.6
234	112.2	283	139.4	332	166.7	381	193.9	430	221.1
235	112.8	284	140.0	333	167.2	382	194.4	431	221.7
236	113.3	285	140.6	334	167.8	383	195.0	432	222.2
237	113.9	286	141.1	335	168.3	384	195.6	433	222.8
238	114.4	287	141.7	336	168.9	385	196.1	434	223.3
239	115.0	288	142.2	337	169.4	386	196.7	435	223.9
240	115.6	289	142.8	338	170.0	387	197.2	436	224.4
241	116.1	290	143.3	339	170.6	388	197.8	437	225.0
242	116.7	291	143.9	340	171.1	389	198.3	438	225.6
243	117.2	292	144.4	341	171.7	390	198.9	439	226.1
244	117.8	293	145.0	342	172.2	391	199.4	440	226.7
245	118.3	294	145.6	343	172.8	392	200.0	441	227.2
246	118.9	295	146.1	344	173.3	393	200.6	442	227.8
247	119.4	296	146.7	345	173.9	394	201.1	443	228.3
248	120.0	297	147.2	346	174.4	395	201.7	444	228.9
249	120.6	298	147.8	347	175.0	396	202.2	445	229.4
250	121.1	299	148.3	348	175.6	397	202.8	446	230.0
251	121.7	300	148.9	349	176.1	398	203.3	447	230.6
252	122.2	301	149.4	350	176.7	399	203.9	448	231.1
253	122.8	302	150.0	351	177.2	400	204.4	449	231.7

Incompatible Chemicals

The substances in the left-hand column must be stored and handled so that they cannot come into any contact with the substances in the right-hand column.

Alakaline and alkaline-earth metals, such as sodium, potassium, cesium, lithium, magnesium, calcium, aluminum	Carbon dioxide, carbon tetrachloride, and other chlorinated hydrocarbons. (Also prohibit water, foam, and dry chemical on fires involving these metals.)
Acetic acid	Chromic acid, nitric acid, hydroxyl-containing compounds, ethylene glycol, perchloric acid, peroxides, and permanganates.
Acetone	Concentrated nitric and sulfuric acid mixtures.
Acetylene	Chlorine, bromine, copper, silver, fluorine, and mercury.
Ammonia (anhydr)	Mercury, chlorine, calcium hypochlorite, iodine, bromine, and hydrogen fluoride.
Ammonium nitrate	Acids, metal powders, flammable liquids, chlorates, nitrites, sulfur, finely divided organics or combustibles.
Aniline	Nitric acid, hydrogen peroxide.
Bromine	Ammonia, acetylene, butadiene, butane and other petroleum gases, sodium carbide, turpentine, benzene, and finely divided metals.
Calcium carbide	Water (See also acetylene.)
Calcium oxide	Water.
Carbon, activated	Calcium hypochlorite.
Copper	Acetylene, hydrogen peroxide.
Chlorates	Ammonium salts, acids, metal powders, sulfur, finely divided organics or combustibles.
Chromic acid	Acetic acid, naphthalene, camphor, glycerol, turpentine, alcohol, and other flammable liquids.
Chlorine	Ammonia, acetylene, butadiene, butane and other petroleum gases, hydrogen, sodium carbide, turpentine, benzene, and finely divided metals.

Chlorine dioxide	Ammonia, methane, phosphine, and hydrogen sulfide.
Fluorine	Isolate from everything.
Hydrocyanic acid	Nitric acid, alkalis.
Hydrogen peroxide	Copper, chromium, iron, most metals or their salts, any flammable liquid, combustible aniline, nitromethane.
Hydrofluoric acid, anhydrous (hydrogen fluoride)	Aqueous or anhydrous ammonia
Hydrogen sulfide	Fuming nitric acid, oxidizing gases.
Hydrocarbons (benzene, butane, propane, gasoline, turpentine, etc.)	Fluorine, chlorine, bromine, chromic acid, sodium peroxide.
Iodine	Acetylene, anhydrous or aqueous ammonia.
Mercury	Acetylene, fulminic acid, ammonia.
Nitric Acid (conc)	Acetic acid, aniline, chromic acid, hydrocyanic acid, hydrogen sulfide, flammable liquids, flammable gases, and nitritable substances.
Nitroparaffins	Inorganic bases.
Oxygen	Oils, grease, hydrogen, flammable liquids, solids or gases.
Oxalic acid	Silver, mercury.
Perchloric acid	Acetic anhydride, bismuth and its alloys, alcohol, paper, wood, grease, oils.
Peroxides, organic	Organic or mineral acids; avoid friction.
Phosphorus (white)	Air, oxygen.
Potassium chlorate	Acids (See also chlorate.)
Potassium perchlorates	Acids (See also perchloric acid.)
Potassium permanganate	Glycerol, ethylene glycol, benzaldehyde, sulfuric acid.
Silver	Acetylene, oxalic acid, tartaric acid, fulminic acid, ammonium compounds.
Sodium	See alkaline metals.
Sodium nitrate	Ammonium nitrate and other ammonium salts.

Sodium oxide	Water.
Sodium peroxide	Any oxidizable substance, such as ethanol, methanol, glacial acetic acid, acetic anhydride, benzaldehyde, carbon disulfide, glycerol, ethylene glycol, ethyl acetate, methyl acetate, and furfural.
Sulfuric acid	Chlorates, perchlorates, permanganates.
Zirconium	Prohibit water, carbon tetrachloride, foam, and dry chemical on zirconium fires.

Safety in the Laboratory or Home Workshop

It is necessary to learn:
 Use of laboratory fume hoods
 Handling flammable solvents
 Mixing acids
 Glass blowing

 Common electrical hazards
 First aid (for four situations only)
  Stoppage of breathing
  Profuse bleeding
  Chemical burns (water only)
  Fire in clothing

 Use of portable fire extinguishers
 Compressed or flammable gases
 Handling and storing dangerous chemicals, including alkali metals.

An outstanding deficiency pertaining to laboratory safety seems to be a lack of awareness of hazards among nontechnical personnel. It is conceivable that increased emphasis on "briefing" custodial workers about the dangers of the laboratories in which they work, and periodic review of these conditions could substantially reduce the hazard of ignorance.

Third, a more universal use of safety glasses, reaction shields, and other personal protective devices seems to be needed. From the responses received, an increased program of education on the hazards of common laboratory procedures and the use of personal protective equipment to lessen these hazards would be helpful.

Chemical Hazards

All laboratories, whether they be biological, chemical, or radiological, utilize hazardous chemicals. The hazard may result from utilizing the "raw" product or from products of a chemical reaction between two or more substances or breakdown products developed through heating or aging. Laboratory personnel should have an acquaintance at least with the modes of entry, the physiological responses, both acute and chronic, and methods of roughly assessing the hazards potential of chemicals they are using.

Electrical Hazards and Management

The problem of handling electricity is probably one of the most ignored facets of safety, yet each year many needless deaths and injuries are caused through carelessness in handling even low voltages. It is also of importance to recognize that electrical equipment can act as an ignition source to activate a fire or explosion. Static electricity should be considered in this category.

Pressure Hazards

Pressure equipment, either high or low (vacuum), is a part of most laboratories. High-pressure apparatus such as gas cylinders, if improperly handled, can be very dangerous. This is especially true of oxygen. Precautions are necessary in handling, transporting, and in storing. Vacuum equipment, through implosions, can be every bit as dangerous as high-pressure explosions.

Cryogenic Hazards

Cryogenics or the use of low-temperature refrigerants requires a knowledge of the behavior of these materials under laboratory conditions. It is impossible to understand the design of a piece of cryogenic equipment or cryogenic experiment without an appreciation for the principles of insulation or the significance of extremely low temperatures. Misuse can result in severe injury.

Flammable Chemicals-Hazards

Fires and explosions account for the most dangerous and the most expensive types of laboratory accidents. A knowledge of the flammable properties of chemicals along with an understanding of potential sources of ignition is extremely vital. Storage and handling of these materials also requires special attention.

General Safety Considerations

A number of accidents and injuries in laboratories could very well result from improper lifting, falls, and lacerations from improper handling of glassware. Preventive measures in these areas are worthy of mention.

Ventilation

The principal method of hazards control in laboratory involves the effective use of ventilation, both general and exhaust. An example of exhaust ventilation is the fume hood which if improperly designed or used fails to give the desired protection. Observations indicate that the function of this equipment is not entirely understood and a number of misuses have been witnessed.

Laboratory Sanitation

Poor laboratory sanitation practices may be the cause of contaminating potable water supplies through temporary cross connections. At times, poor housekeeping practices may be dangerous because of blocking passages or by providing tripping hazards. Many chemicals are kept well beyond their usefulness, causing containers to deteriorate and leak, or chemicals to become unstable. Disposal of flammable and toxic chemicals also presents a problem.

Protective Equipment

All laboratories require that protective equipment of one type or another be immediately available. These devices may include eye wash, emergency shower, safety glasses, eye shields, protective clothing, and respiratory protection. Knowledge of the proper usage and limitations of such equipment is extremely important. At times, injury or death may result from improper selection and application of protective equipment.

Reports and Records

Reports and records are necessary adjuncts to any safety program and should be complete, accurate, and disseminated to the appropriate administrators. Accident reports are of little value unless periodically examined and tabulated in order to obtain a picture of local and overall problems.

Emergencies

The initial procedures one follows in an emergency oftentimes determine the ultimate outcome of the accident, both to the individuals and to the installation. The rudiments of first aid, fire fighting and reporting are vital. Personnel have to be continually instructed on procedures for medical and file emergencies and how and where to make these initial contacts. Such procedures are critical, especially when working alone.

Contact Lenses

It is important to wear eye protection in the chemical laboratory and expecially when wearing contact lenses.

Danger in Handling Acid

The heat evolution of the solution could have provided sufficient thermal shock to the glass to permit it to crack when lifted free of the counter, or setting it on a cold counter top; or a shock in setting it down, could have contributed to the bottom separating when the jug was lifted.

Written procedures for handling of acids should always be followed. Personal protective equipment consisting of face protection, rubber apron, and gloves are a necessity for this operation.

You Must Have Fire Extinguishers

If a fire breaks out in your office or apartment, get out fast. Many people are killed because they don't realize how fast a small fire can spread.

If you are caught in smoke take short breaths, breathe through your nose, and crawl to escape. The air is better near the floor.

Head for stairs — not elevator. A bad fire can cut off the power to elevators. Close all doors and windows behind you.

If you are trapped in a smoke-filled room, stay near the floor, where the air is better. If possible, sit by a window where you can call for help.

Feel every door with your hand. If it's hot don't open. If it's cool, make this test: open slowly and stay behind the door. If you feel heat or pressure coming through the open door, slam it shut.

If you can't get out, stay behind a closed door. Any door serves as a shield. Pick a room with a window. Open the window at the top and bottom. Heat and smoke will go out the top. You can breathe out the bottom.

DON'T fight a fire yourself.

DON'T jump. Many people have jumped and died without realizing rescue was just a few minutes away.

If there is a panic for the main exit, get away from the mob. Try to find another way out. Once you are safely out, *DON'T* go back in. Call the Fire Department immediately. Use alarm box or telephone. DIAL 911

If you find smoke in an open stairway or open hall, use another pre-planned way out.

REMEMBER: Get out fast. Don't underestimate how fast a small fire can spread. Use stairs, not the elevator. Close all doors behind you. Don't panic. Once you are safely out, call the Fire Department. Dial 911 or use alarm box. Don't go back in.

TRADEMARK CHEMICALS

The numbers to the right of each tradename product refers to the suppliers who are listed immediately following this trademark list.

AC-629	5	Antarox®	34
Accoquat	13	Aromox	12
Accosperse	13	Aromatic 150	31
Acetulan	7	Arquad®	12
Acme Oil B	76	Arylan	51
Adogen	14	Biopal	34
Adol	10	Bio Soft	78
Alfonic	23	Bio Terge	78
Alkamide	4	Blancophor®	34
Alkamidox	4	Bronopol	1
Alkamox	4	Cab-O-Sil	18
Alkaphos	4	Carbitol	81
Alkaquat	4	Carbowax	81
Alkasurf	4	Carnauba Spray 200	46
Alkateric	4	Carnauba Wax	46
Alkatrope	4	Castor Oil	16
Alkawet	52	CDB-69	33
Alox	6	Ceraphyl	82
Amerchol	7	Cerasynt	72
Amerlate	7	Cetab	69
Amidox	78	Cetrimide	20

Chlorothene 28
Clindrol. 22
CMC 7M 41
Conco 25
Condensate CO 24
Condensate PS 24
Crill 27
Crillet 27
Crillon. 27
Crodapearl 26
Crodapur Lanolin 26
Crodaquest. 27
Crodasinic 26
Cyclo Sol®. 73
Dowanol 28
Dowicide. 28
Dowicil 28
Durez 42
Emerest. 29
Emiphos 84
Empicol. 80
Empilan. 2
ESI-Terge. 30
Ethofat® 12
Ethomeen® 12
Eumulgin®. 40
Foamole 82
G-1292 15
Gafac 34
Gantrez. 34
Gloquat. 37
Glucam 7
Hamposyl 39
Hartolan 26
Heliogen® 35
Hoescht Wax 8
Hydrofol 10
Igepal 34
Igepon. 34
Kaopolite. 49
Kasil 66
Kelzan. 50
Kessco. 12
Kesscolin 12
Klearfac. 17
Klucel 41
Lanacet 53
Latex E-295 60
Lathanol 78
Lexaine 47
Lexein. 47
Lexemul 47
Lexol 47
Maprofix 68
Macol 55
Makon. 78
Monaquest 58
Maphos 55
Mazon. 55
Methocel 28
Metso 66
Miranol® 57
Modulan 7
Monacor 58
Monofax 58
Monamate 58
Monamid 58
Monamine 58
Monamulse. 58
Monaterge 58
Monateric 58
Monawet 58
Monazoline 58
Myrj 45
N Sodium Silicate 66
Nacap 83
Nacconol 78

SUPPLIERS OF TRADEMARK CHEMICALS

1. Air Products and Chemicals Inc. . . . . .Allentown, Pa.
2. Albright & Wilson, Marchon Div. . . . Whitehaven, Cumbria, England
3. Alcolac Inc. . . . . .Baltimore, Md.
4. Alkaril Chemicals Ltd. . . . . .Mississauga, Ont., Canada
5. Allied Chem. Corp., Semet Solvay Div. . . . . .New York, N.Y.
6. Alox Corp. . . . . Niagara Falls, N.Y.
7. Amerchol Corp. (Unit of CPC Internat'l. Inc.). . . . . . Edison, N.J.
8. American Hoescht Corp. . . . . Somerville, N.J.
9. Antara Chemicals Div. . . . . N.Y., N.Y.
10. Archer-Daniels-Midland Co. . . . . .Minneapolis, Minn.
11. Arco Chem. Div., Atlantic Richfield Co. . . . . .Philadelphia, Pa.
12. Armak Industrial Chemicals. . . . . .Chicago, Ill.
13. Armstrong Chem. Corp. . . . . Janesville, Wisc.
14. Ashland Chem. Co. . . . . Dublin, Ohio
15. Atlas Chem. Industries (see ICI United States) . . . Wilmington, Del.
16. Baker Castor Oil Co. . . . . Bayonne, N.J.
17. BASF Wyandotte Corp., Industrial Chemicals Group.
    . . . . .Wyandotte, Mich.
18. Cabot Corp. . . . . Boston, Mass.
19. Chemco Products Co. . . . . .Chicago, Ill.
20. Chemo Puro Manufacturing Co. . . . . Long Island City, N.Y.
21. Ciba-Geigy Corp., Dyestuff & Chemicals Div. . . . . Greensboro, N.C.
22. Clintwood Chem. Co. . . . . .Chicago, Ill.
23. Conoco Chemicals, Continental Oil Co. . . . . Houston, Texas
24. Continental Chem. Co. . . . . Clifton, N.J.
25. Continental Oil Co. (see Conoco Chemicals) . . . . . . Houston, Texas
26. Croda Inc. . . . . .New York, N.Y.
27. Croda Chem. Ltd. . . . . .Surrey, England
28. Dow Chemical Corp. . . . . .Midland, Mich.
29. Emery Industries, Malmstrom Chemicals . . . . Linden, N.J.
30. Emulsion Systems, Inc. . . . . Brooklyn, N.Y.
31. Exxon Company . . . . Houston, Texas

32. Falleck Chem. Corp. . . . . .Newark, N.J.
33. FMC Corp. . . . . .New York, N.Y.
34. GAF Chem. Prod. Div. . . . . .New York, N.Y.
35. General Dyestuff Corp. . . . . .New York, N.Y.
36. General Electric . . . . .Waterford, N.Y.
37. Glovers Chem. Ltd. . . . . Leeds, England
38. Goldschmidt Prod. Corp. . . . . .White Plains, N.J.
39. W.R. Grace & Co., Organic Chem. Div. . . . . Lexington, Mass.
40. Henkel Chemicals (Canada) Ltd.. . . . . (Montreal), P.Q. Canada
40A. Henkel Inc. Chemical Specialties Div. . . . . .Hoboken, N.J.
40B. Henkel KGaA . . . . .Dusseldorf, West Germany
41. Hercules Inc., Organic Dept. . . . . Wilmington, Del.
42. Hooker Chem. Corp., Durez Plastic Div.. . . North Tonawanda, N.Y.
43. Hostawax Div. . . . . .New York, N.Y.
44. Humble Oil and Refining Co.. . . . . Humble, Tx.
45. ICI USA Inc. . . . . Wilmington, Del.
46. Innis, Speiden and Co. . . . . .New York, N.Y.
47. Inolex Corp., Personal Care Div.. . . . . .Chicago, Ill.
48. Johns-Manville Product Corp. . . . . .New York, N.Y.
49. Kaopolite Co. . . . . .Elizabeth, N.J.
50. Kelco Co.. . . . . San Diego, Calif.
51. Lankro Chem. Ltd. . . . . .Manchester M30 OBH, England
52. Lonza Inc. . . . . Fair Lawn, N.J.
53. Malmstrom Chem. Corp. Div. of Emery Industries
. . . . Linden, N.J.
54. A. R. Mass Chem. Co.. . . . . .South Gate, Calif.
55. Mazer Chem., Inc. . . . . .Gurnee, Ill.
56. McLaughlin Gormley King Co.. . . . . .Minneapolis, Minn.
57. Miranol Chem. Co.. . . . . Irvington, N.J.
58. Mona Industries . . . . Paterson, N.J.
59. Monsanto Co. . . . . St. Louis, Mo.
60. Morton Chem.–Div. of Morton-Norwich Prod. Inc.
. . . . Woodstock, Ill.
61. National Starch & Chem. Corp. . . . . .New York, N.Y.
62. NL Industries, Baroid Div.. . . . . Houston, Texas
63. Ottawa Chem.. . . . . .Toledo, Ohio
64. Ozokerite Mining Co. . . . . .Grand Rapids, Mich.
65. Patco Coating Products, C.J. Patterson . . . . Kansas City, Mo.
66. Philadelphia Quartz Co. . . . . .Philadelphia, Pa.

67. Polyvinyl Chemicals....................Wilmington, Ma.
68. Refined-Onyx Div., Millmaster Onyx Corp........Lyndhurst, N.J.
69. Rhodes Chemical Corp...................Jenkintown, Pa.
70. Rohm and Haas Co......................Philadelphia, Pa.
71. Schenectady Chemicals.................Schenectady, N.Y.
72. Schylkill Chem. Co.....................Philadelphia, Pa.
73. Shell Chem. Co........................Houston, Texas
74. Sonneborn Sons, Inc....................New York, N.Y.
75. Standard Dry Wall Prod...................New Eagle, Pa.
76. Standard Oil Co........................Chicago, Ill.
77. Stauffer Chem. Co......................Westport, Conn.
78. Stepan Chem. Co.......................Northfield, Ill.
79. Textilana Corp., Div. of Henkel Inc..........Hawthorne, Calif.
80. Uceto Chem. Co........................Flushing, N.Y.
81. Union Carbide Corp., Chemicals and Plastics......New York, N.Y.
82. Van Dyke & Co., Inc....................Belleville, N.J.
83. R. T. Vanderbilt.......................Norwalk, Conn.
84. Witco Chem., Organic Div................New York, N.Y.
85. Witco Chem., Ultra Div..................Paterson, N.J.
86. Witco Chem. Ltd......................Worcester, England

INDEX